Construction
Program
Management

Best Practices and Advances in Program Management Series

Series Editor
Ginger Levin

PUBLISHED TITLES

Construction Program Management
Joseph Delaney

Applying Guiding Principles of Effective Program Delivery
Kerry R. Wills

Program Management: A Life Cycle Approach
Ginger Levin

*Implementing Program Management: Templates and Forms Aligned
with the Standard for Program Management,
Third Edition* (2013) *and Other Best Practices*
Ginger Levin and Allen R. Green

FORTHCOMING TITLES

Program Governance
Muhammad Ehsan Khan

*Successful Program Management:
Complexity Theory, Communication, and Leadership*
Wanda Curlee and Robert Lee Gordon

Sustainable Program Management
Gregory T. Haugan

The Essential Program Management Office
Gary Hamilton

*Leading Virtual Project Teams: Adapting Leadership Theories
and Communications Techniques to 21st Century Organizations*
Margaret R. Lee

From Projects to Programs: A Project Manager's Journey
Samir Penkar

Construction
Program
Management

Joseph Delaney

CRC Press
Taylor & Francis Group
Boca Raton   London   New York

CRC Press is an imprint of the
Taylor & Francis Group, an **informa** business
AN AUERBACH BOOK

CRC Press
Taylor & Francis Group
6000 Broken Sound Parkway NW, Suite 300
Boca Raton, FL 33487-2742

© 2014 by Taylor & Francis Group, LLC
CRC Press is an imprint of Taylor & Francis Group, an Informa business

No claim to original U.S. Government works

Printed on acid-free paper
Version Date: 20130514

International Standard Book Number-13: 978-1-4665-7504-2 (Hardback)

Visit the Taylor & Francis Web site at
http://www.taylorandfrancis.com

and the CRC Press Web site at
http://www.crcpress.com

To my father, David Dennis Delaney, who was my inspiration and motivation for pursuing a career as a professional engineer

Contents

Preface

Leadership is the capacity to translate vision into reality.

Warren G. Bennis

With those words on leadership, Warren G. Bennis, fellow State University of New York (SUNY) at Buffalo alumnus, and a pioneer in the field of leadership studies, could have clearly been defining the role of a modern-day construction program manager. Transforming vision into reality is really a twofold process in construction program management. First there is the technical challenge of translating from the ideal world of plans and specifications into the real world of steel and concrete. Second is the leadership challenge of aligning and then focusing the team on the mission, vision, objectives, and strategy to achieve the desired program outcome and then to realize its anticipated benefits. The construction program manager must lead the team to succeed in both the technical and leadership challenges for a program to be considered a true success. This will require both tactical and strategic thinking. It will take a unique individual with a combination of experience, technical know-how, emotional intelligence, and personal ethics.

This book presents a process-based approach to construction program management. The model presents leveraged structures to bring order to what can otherwise feel like an overwhelming challenge. As the case studies show, with the right team and leader, and with the proper implementation of the steps outlined within, all programs can obtain true success. I guarantee it.

Acknowledgments

This book holds many memories and contains the knowledge earned from a rewarding career as a construction professional. The fondest memories are the remarkable people who not only shaped my professional life, but also helped me develop into a better person. Over the past 25 years I have accumulated many debts of gratitude, a few of which I have space to acknowledge here.

First I would like to acknowledge the excellent faculty at the State University of New York (SUNY) at Buffalo who supported me in my pursuit to become a professional engineer. I would like to specially thank Dr. Satish Mohan, who as my leading professor, pointed me toward the construction path, developing the Master of Engineering in Construction Management Program, of which I became the first graduate.

After graduate school I was fortunate to start my career at Peter Kiewit Sons, Inc. Kiewit is an outstanding company. They do it right and never cut corners. And they do it with integrity and never compromise ethics. Kiewit was my baptism by fire, where I developed construction know-how, unfortunately sometimes through the school of hard knocks. I would specially like to acknowledge Butch Shaver and Kirk Sameulson who guided me through that process, and thank them for their patience and support.

For the second part of my career I joined the "dark side" and became a professional consultant. These were my years at the C&S Companies, where I started as a CM and was ultimately given the extraordinary opportunity to run C&S Design Build, Inc., one of four C&S Companies. There are many folks at C&S who helped get me there, too many to name in fact, several who are now personal friends. I would specially like to acknowledge John Trimble and Mike Walker, who were always there for me, through thick and thin.

BOOK REVIEW TEAM

I was very fortunate to assemble a "dream team" of project and program professionals to review this book. This diverse group of individuals provided a fresh set of eyes, and very often unique perspectives, on the book's concepts, deductions, and conclusions. Their efforts are greatly appreciated. Their suggestions and contributions made this a better book. The team consisted of the following individuals.

Dr. Ginger Levin is a senior consultant and educator in project management. Her specialty areas are portfolio management, program management, and the project management office, metrics, and maturity assessments. She is certified as a PMP, PgMP, and as an *OPM3* Certified Professional. She was the second person in the world to receive the PgMP. As an *OPM3* Certified Professional, she has conducted over 20 maturity assessments using the *OPM3* Product Suite tool.

In addition, Dr. Levin is an adjunct professor at the University of Wisconsin–Platteville where she teaches in its MS project management program, and for SKEMA (formerly Esc Lille) University, France, and RMIT in Melbourne, Australia in their doctoral programs in project management.

In consulting, she has served as project manager in numerous efforts for Fortune 500 and public sector clients, including Genentech, Cargill, Abbott Vascular, UPS, Citibank, the US Food and Drug Administration, General Electric, SAP, EADS, John Deere, Schreiber Foods, TRW, New York City Transit Authority, the US Joint Forces Command, and the US Department of Agriculture. Prior to her work in consulting, she held positions of increasing responsibility with the US government, including the Federal Aviation Administration, Office of Personnel Management, and the General Accounting Office.

Dr. Levin is the author of *Interpersonal Skills for Portfolio, Program, and Project Managers*, published in 2010. She is the co-author of *Program Management Complexity: A Competency Model* (2011), *Implementing Program Management: Forms and Templates Aligned with the Standard for Program Management*, second edition (2008), *Project Portfolio Management*, *Metrics for Project Management*, *Achieving Project Management Success with Virtual Teams*, *Advanced Project Management Office: A Comprehensive Look at Function and Implementation*, *People Skills for Project Managers*, *Essential People Skills for Project Managers*, *The Business Development Capability Maturity Model*, and ESI's *PMP Challenge! PMP Study Guide* and the *PgMP Study Guide*.

Dr. Levin received her doctorate in information systems technology and public administration from The George Washington University, and received the *Outstanding Dissertation Award* for her research on large organizations.

David T. Boyle obtained his doctorate in educational administration from SUNY–Buffalo. In addition he holds two Master of Education degrees, one in English and a second in reading. His undergraduate studies were completed at SUNY–Fredonia. Most recently, he retired from the Cayuga-Onondaga BOCES after 23 years as the assistant superintendent for finance and management services. Previous experience includes employment at the former Genesee–Wyoming BOCES (now called Genesee Valley BOCES) along with various jobs in education including English teacher, reading teacher, educational consultant, and adjunct professor at SUNY–Cortland.

Dave volunteers as the treasurer of East Hill Family Medical in Auburn, New York. He has always enjoyed the arts and serves on the Genesee Symphony Orchestra Board of Directors as its treasurer. In addition he continues his association with BOCES as the citizen member of the Genesee Valley BOCES Audit Committee. Among his many accomplishments, he includes the successful construction of the $43 million Cayuga-Onondaga main campus in Auburn, New York, the first K–12 public school building in New York State awarded status as a LEED building by the US Green Building Council.

Eric Robert serves as dean of the Applied Sciences & Building Technologies Division at SUNY–Delhi, where he oversees academic programs in the areas of natural resource recreation and sports, golf and plant sciences, architecture, construction technology and management, integrated energy systems, electrical instrumentation and controls, and several other related applied offerings. During his prior role as a full-time faculty member within the college's construction department, Eric taught several courses focusing on the legal and managerial aspects of the industry and also technical offerings related to mechanical and environmental systems in various settings. Eric served in a lead role toward expanding on the college's strong foundation and developing Delhi's own construction management (BT) degree, which was first offered in 2007 and has quickly become one of the highest regarded programs of its kind in the Northeast.

In addition to his academic career, Eric has extensive experience managing a variety of construction projects. While with Whiting-Turner, he served as a project manager on a large microelectronics program for International Business Machines (IBM) (Fishkill, NY), which collectively reached over $1 billion in completed construction, along with other major capital undertakings at Lehigh Valley Hospital's Cedar Crest Campus (Allentown, PA) and Bryn Athyn College (Bryn Athyn, PA). Currently, Eric serves as a consulting project manager and advisor with the C&S Companies where he has managed several K–12 school projects and a mid-rise condominium development in downtown Albany, New York.

Eric received his associates degree in applied sciences in construction technology from SUNY–Delhi in 1996. He later continued his studies in construction management at Purdue University, where he received a bachelor's of science, and at SUNY Environmental School of Science & Forestry, where he completed a masters of professional studies. Eric also obtained a master's in business administration from Mount Saint Mary College. He recently completed a certificate in advanced studies in higher education administration from SUNY–Albany.

Charles A. Bouley, Jr. (Chuck) is president of Bouley Associates, Inc., a general contractor located in Auburn, New York. Bouley Associates is a third-generation family business which, since the early 1950s, has constructed many of the most significant projects in upstate New York. Chuck graduated from Syracuse University with a bachelor of science degree in construction management in 1980. In addition to his 20-plus years in the family business he worked as a project manager for more than 10 years for a large mechanical contractor building wastewater treatment facilities, up to $50 million in contract value.

Samuel J. Cichello (Sam) is the owner of the Samuel J. Cichello, Architect firm in Weedsport, New York, established in 1963. Sam is a licensed architect in New York State, and was formerly licensed in Pennsylvania, New Jersey, and Maryland. He graduated from Syracuse University in 1954 and throughout the years has served as a visiting lecturer there. He was a consultant for the NYS Educational Facilities Planning publication, *Thermal Environment of Educational Facilities*. He received an AIA National Award of Merit in recognition for an outstanding contribution to "Homes for Better Living" (shared with William H. Scarborough, AIA). Sam also was the lead architect for the $42 million Cayuga-Onondaga BOCES campus in Auburn, New York which received a silver award for the first LEED-certified public school facility in New York State.

I would also like to give special thanks to **Mary Kathleen Delaney** who spent many hours reviewing the manuscript and providing suggestions and comments. Her unique perspective as both my wife and a former project coordinator was very valuable and much appreciated. Mary Kathleen is currently the owner of Delaney CMS, a construction management and real-estate company located in Union Springs, New York.

About the Author

Joseph Delaney is president and general manager of the construction management services group at Delaney CMS in Union Springs, New York. He is the former president of C&S Design Build Inc., C&S's program and construction management company, where he started the agency construction management business, and in his 12-year career there, developed it into the fastest-growing and most profitable of the firm's business units. At C&S, Joe was responsible for more than $1 billion in  completed projects in the education, municipal, wastewater, airport, and private development sectors. In his 25 years as a construction professional, Joe has served in many roles including those of project engineer, design coordinator, superintendent, construction manager, program manager, general manager, and president. Joe is a licensed professional engineer, holds the Program Management Professional (PgMP®) credential, is a Certified Construction Manager (CCM®), and is a Leadership in Energy and Environmental Design (LEED®) accredited professional. Joe is the only person in the world with all of these credentials. In addition, he is an expert critical path method scheduler and has presented both nationally and internationally on construction program management. Joe earned his bachelor's and master's degrees in civil engineering from the State University of New York at Buffalo. He enjoys sharing his experience leading teams, and has provided leadership training to senior management including his colleagues at the American Council of Engineering Companies (ACEC) Senior Executive Institute and at Fortune 500 companies, including Johnson & Johnson, Inc. Joe was recently the keynote speaker at the Tri-State Diversity Summit where he shared Delaney CMS's (a woman-owned business enterprise) vision for equality and empowerment.

1

Process-Based Management Approach

1.1 INTRODUCTION

The construction industry is one of the largest enterprises in the United States. According to the US Department of Commerce, total construction spending in the United States totaled $9.45 trillion in 2011 alone. However, the US construction industry lags behind other industries in its implementation of modern-day management techniques. Widely accepted management principles, such as those contained in the *Standard for Program Management* (the *Standard*) by the Project Management Institute® (PMI) are not widely understood nor implemented. This book explores how improved understanding and implementation of the *Standard*, with a few tweaks for construction programs, could improve success rates.

This opening chapter focuses on basic definitions of project management, program management, and explores similarities and differences. A summary review of the *Standard* focuses on how these management concepts can be applied to capital construction programs. We then drill down and explore the implementation of the *Standard* throughout the program management life-cycle phases (initiation–planning–execution–monitor/control–closure) in future chapters.

1.2 PROJECT MANAGEMENT

When I started the construction management business at the C&S Companies, I was fortunate to be able to initiate this new business endeavor from scratch. The C&S Companies was an Engineering New Record Top 200® engineering company at the time, but this was its first attempt at developing a construction management business unit. To its credit, it realized that simply rebranding the engineering business would not work and that an entirely different approach was needed. What I saw was a unique opportunity to do it right and not be handcuffed by legacy processes and procedures.

Our first project, as a separate business unit, was a $132 million Metropolitan Wastewater Treatment Plant (the Metro project). This was part of a $650 million program to reduce pollutants entering Onondaga Lake (the Lake Improvement Program Office–LIPO program). The program included all of Onondaga County's treatment and distribution systems in the Syracuse, New York area, which in total discharge over 240 million gallons per day of effluent into Onondaga Lake and its tributaries. Talk about a baptism by fire!

Probably because of my training as a professional engineer, I realized that a structured, and process-based, management approach was needed for such a large and complex project and program. My first order of business was to research current best practices for construction and program management. Interestingly enough, I found very little useful data on construction and program management but quite a bit of information on project management. This is where I was introduced to PMI® and what I consider a very well-developed process for project management, the *Guide to the Project Management Body of Knowledge (PMBOK guide)* [1]. PMI was established in 1969 and has grown today into a worldwide organization that advocates for the project management profession and sets professional standards for project and program managers. In 1984 PMI established a unique certification program for project managers called the Project Management Professional (PMP®) credential. Since the start of the Metro project, this credential has grown exponentially, from fewer than 25,000 individuals back then to over 500,000 today.

So when we started the Metro project, the trick was to see if the PMI concepts for project management could be applied to large and complex capital projects. The basic premise of the *PMBOK* guide is that all projects are similar in nature. Any project can be described as a temporary endeavor undertaken to create a unique product or service [1]. Projects have a fixed start and end (temporary) and they are executed to produce something. So whether it is software development, drug discovery and delivery, design and production of an automobile, even baking a birthday cake, and presumably building a wastewater treatment plant, the PMI approach would apply. Another basic concept of the *PMBOK* guide is that all projects can be successfully accomplished by implementing a standard set of management processes. This approach is consistent with other management standards such as ISO 9000 and the Software Engineering Institute's CMMI. As I was looking for a structured management approach for the Metro project, this appeared to be just the right fit. So we went with it.

To better appreciate the PMI approach, it is important to recognize what a process is. A process can be defined as a series of actions or steps taken to achieve an end [2]. More precisely, a process is a sequence of interdependent and linked procedures that, at every stage, consume one or more resources to convert inputs into outputs [3]. These outputs then serve as inputs for the next stage until the known goal or end result is reached [3]. The *PMBOK* guide describes the project management process in a similar way, as:

- Inputs (documents, plans, designs, etc.)
- Tools and techniques (mechanisms applied to inputs)
- Outputs (documents, products, etc.)

The *PMBOK* guide recognizes five basic project management process groups that are organized in a framework of nine knowledge areas. The five basic project management process groups are defined as follows:

- *Initiating Process Group:* Those processes performed to define and authorize a new project
- *Planning Process Group:* Those processes required to establish the scope of the project, refine the objectives, and define the course of action required to attain the objectives that the project was undertaken to achieve
- *Executing Process Group:* Those processes performed to complete the work to satisfy the project specifications
- *Monitoring and Controlling Process Group:* Those processes required to track, review, and regulate the progress and performance of the project; identify any areas in which changes to the plan are required; and initiate the corresponding changes
- *Closing Process Group:* Those processes performed to finalize all tasks and to formally close the project

The nine knowledge area definitions are:

- *Scope Management:* A set of processes used to draw boundaries around the project by specifying what is included and what is not

- *Time Management:* Ensures that the project is completed on schedule
- *Cost Management:* Ensures that the project is completed on budget
- *Quality Management:* Ensures that the project meets its quality requirements and performs as planned
- *Human Resource Management:* All of the processes used to develop, manage, and put the project team together
- *Communication Management:* Determines what information is needed, how that information will be sent and managed, and how project performance will be reported
- *Risk Management:* Processes to identify, manage, and control risk and exploit opportunities
- *Procurement Management:* A group of processes used to acquire the materials and services needed to complete the project
- *Project Integration Management:* Coordinates the other knowledge areas to work together throughout the project

For me, it is easiest to look at the process groups as which work is to be done when and the knowledge areas as specialized expertise on how to get the work done. There are 44 distinct project management processes, each one having specific inputs, techniques, and outputs. Each process belongs to a process group and a specific knowledge area. The resulting ingenious arrangement is a complete outline of all the required tasks and techniques to manage any type of project. Each of the nine knowledge areas can be mapped to the five processes as shown in Table 1.1 [1].

We investigate each of the knowledge areas and processes in much more detail in Chapter 2. As far as the Metro project goes, the *PMBOK* guide approach worked very well. We completed the project ahead of schedule, under budget, and the facility functioned flawlessly from start-up. The approach can be credited with providing a detailed road map that kept us focused on critical project management tasks.

1.3 PROGRAM MANAGEMENT

When we initiated the Metro project in early 2000 there were no universal standards for program management [4]. What I found was that the term was being used interchangeably to describe different things. Of course this was not helpful in trying to establish a standard management process for

TABLE 1.1

Knowledge Areas and Process Groups

Knowledge Areas		Project Management Processes			
	Initiate	Plan	Execute	Monitor and Control	Close
Scope Management	—	Set requirements	—	Verify scope	—
		Define project scope	—		—
		Create WBS	—	Change control	—
Time Management	—	Define tasks	—	Monitor schedule	—
		Sequence tasks	—		—
		Assign resources	—		—
		Develop schedule	—	Control schedule	—
Cost Management	—	Resource planning	—	Monitor budget	—
		Estimate costs	—		—
		Establish budget	—	Control cost	—
Quality Management	—	Quality management planning	Quality assurance	Quality control	—
Human Resource Management	Identify Stakeholders	Human resource planning	—	—	—
		Organizational planning	Develop project team	—	—
		Staff acquisition	—		—
Communication Management	—	Communication planning	Distribute information	Report performance	—

(Continued)

TABLE 1.1 (*Continued*)
Knowledge Areas and Process Groups

Knowledge Areas		Project Management Processes			
	Initiate	Plan	Execute	Monitor and Control	Close
Risk Management	—	Identify risks	—	Monitor risk	—
		Qualitative risk analysis	—		—
		Quantitative risk analysis	—	Control risk	—
		Plan risk responses	—	—	—
Procurement Management	—	Procurement planning	Vendor solicitation	Monitor contracts	Contract closure
			Vendor selection	Control contracts	
Project Integration Management	Develop charter	Develop project plan	Execute the work	Monitor tasks	Close project
				Control tasks	
				Change control	

Metro and the LIPO program. I knew at the time that managing a program with multiple projects was different in many ways from managing a single project.

Today there is much more information available on standard practices for program management. For capital programs, the best resources are PMI and the Construction Management Association of America (CMAA). The CMAA is a professional association formed in 1982 that serves the construction management industry. The CMAA is much smaller than the PMI, with current membership at approximately 10,000 [5]. Membership includes individual CM/PM practitioners, corporate members, and owners in both the public and private construction markets. The CMAA is also an advocacy group that represents the construction management industry before the US Congress, federal agencies, state and local governments, and industry stakeholders. Similar to PMI's certification programs and standards for project managers, the CMAA established a standard of practice and certification program for construction managers, known as the Certified Construction Manager (CCM) credential in 1986. Today there are nearly 2,000 CCMs worldwide.

Although the PMI and the CMAA associations represent similar professions, they define program management in very different ways. This can lead to confusion, particularly with construction professionals familiar with, and who want to leverage, standards of practice. The competing definitions are as follows:

- *PMI:* "Program Management is the management of a set of related projects in a coordinated fashion to obtain control and benefits that would not be available if the projects were managed individually. . . . Like projects, programs are a means of achieving organizational goals and objectives, often in the context of a strategic plan." [6]
- *CMAA:* "Program Management is the practice of professional construction management applied to a capital improvement program on one or more projects from inception to completion. Comprehensive construction management services are used to integrate the different facets of the construction process—planning, design, procurement, construction and activation—for the purpose of providing standardized technical and management expertise on each project." [7]

For me, the best way to look at PMI's definition is that projects deliver outputs, whereas programs deliver outcomes. If we use this distinction and apply it to both the Metro project and the LIPO program as examples, the project delivered a wastewater treatment facility, and the program created a clean Onondaga Lake. In contrast, under CMAA's definition, program management is a process to leverage standardization and to provide for a more collaborative and integrated way to manage either a single project or groups of projects.

Aspects of both definitions apply to the more specific process of construction program management. I would describe program management for construction professionals as follows. *Construction program management* is the practice of incorporating both strategic and tactical management techniques, on one or more projects, from inception to completion, to increase the likelihood of success.

This aligns closely with PMI's definition and includes the value-added component of CMAA's definition. It is important under my definition of construction program management to understand fully the difference between strategic and tactical management techniques. This is, in my mind, the primary difference between project management and program management. Table 1.2 contrasts the two.

Program management requires a higher level of thinking than project management. Unfortunately, my experience is that most construction managers are very good at tactical skills but lack the skills required for strategic planning and struggle as program managers. This is similar to other professions, such as engineering, where technical skills are easily taught and developed, whereas leadership ability is not.

Probably the best way to appreciate strategic thinking is to look at the most famous example of them all: the Apollo program to put a man on the moon. It all started in 1961 with a speech by President John F. Kennedy (JFK) to the US Congress which set out the following bold mission and vision: "I believe

TABLE 1.2

Tactical versus Strategic Thinking

Tactical	Strategic
Project management	Program management
Doing things right	Doing the right things
Doing	Planning
How	Why
Management	Leadership
Narrow view	Broad view
Short term	Long term

that this nation should commit itself to achieving the goal, before this decade is out, of landing a man on the Moon and returning him safely to the Earth."

All programs should be initiated with a powerful mission and vision statement like that! The mission statement establishes what the ultimate purpose is, and the vision statement sets what success will look like. Look at both as rallying points for the program team. I think you will agree that JFK did a pretty good job at that.

Strategic thinking also involves developing a high-level roadmap that frames other components of governance to ensure alignment with the program's mission and vision statements. Governance comes from the Latin word "to steer," so it includes defining requirements and the setting of priorities. But in my mind, even more important, proper governance determines the program's feasibility and the readiness of the team to execute it.

For a construction program governance is especially important. The Central Artery/Tunnel Program (the Big Dig) in Boston is a great example of the negative consequences of proper governance not taking place. The Big Dig is infamous as the most expensive highway program in the United States, plagued by escalating costs, scheduling overruns, leaks, design flaws, charges of poor management, the use of substandard materials, and criminal arrests [8]. One interesting fact that is not well known is that the original estimated program cost ($2.8 billion) was ultimately not even enough to pay for the management consultants assigned to the program. By some estimates, the program will ultimately cost the Commonwealth of Massachusetts a total of $22 billion [9]. The original schedule for the project was seven years. Actual completion took 17 years. I would argue that the feasibility of the program and the readiness of the team were not properly vetted in 1987 when the program was initiated.

In the summer of 1992, while I was working for Peter Kiewit Sons, Inc., I helped develop the CPM master schedule for the South Boston Approach to the Ted Williams Tunnel, one of the very first projects (of 109) in the program. I can still remember the day when our team presented the CPM schedule to the owner. Our construction trailer was literally overflowing with the owner's staff, consulting engineers, and program managers. You can envision our apprehension; five of us from the contractor's team sitting across the conference table from all of them. I guess we now know where some of that $2.8 billion went! As a management consultant myself these days, I generally do not like to be critical of management teams, but

TABLE 1.3

Strategic and Tactical Thinking

		Strategy	
		Doing the Right Things	**Doing the Wrong Things**
TACTICS	Doing Things Right	Success (full alignment)	Quick failure
	Doing Things Poorly	Endure (partial alignment)	Slow failure

I think looking back that you could argue that these folks were not focusing on the right things, especially early in this program. I think they may have completely skipped the program initiation process.

Successful program management involves tactical thinking as well. Probably the best way to look at it is that tactical plans represent the short-term efforts to achieve the strategic, longer-term goals. I am sure that sometime in your professional career you were warned, "Do not miss the forest for the trees," or something similar. You probably also have been told to "Keep your nose to the grindstone," or something to that effect. In program management you really need to do both simultaneously. No easy task. Tactics and strategy must be in alignment, as Table 1.3 illustrates.

1.4 INTRODUCTION TO THE *STANDARD* FOR PROGRAM MANAGEMENT BY PMI

In my opinion, the best standard of practice for program management was, and continues to be, developed by the PMI. You might think that I am biased as one of few PgMPs around but I hold the CCM credential as well. What I like about the *Standard* is that it is a process-based management approach, similar to the *PMBOK* guide, but it clearly distinguishes program management from project management. The *Standard* was first published in 2006, and the current edition in 2013.

1.4.1 The Three Themes of Program Management

Construction professionals are familiar with project management being portrayed as the delicate balance between the triple constraints

FIGURE 1.1
Project management triple constraints.

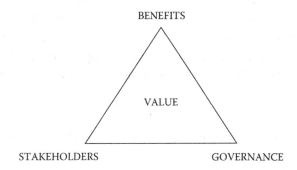

FIGURE 1.2
The three themes of program management.

of scope, cost, and schedule and the ultimate level of quality as illustrated in Figure 1.1. In a similar fashion, the *Standard* describes program management as the delicate balance between the three themes of stakeholder management, benefit realization, and governance and the ultimate level of program value (see Figure 1.2). For better recall, we can illustrate this in a manner analogous to the triangle graphic in Figure 1.1.

We have already discussed governance, so that leaves the themes of stakeholder management and benefit realization to explore. We start with stakeholder management. A *stakeholder* is an entity or individual who has something to gain or lose from the implementation of any phase of the program. *Stakeholder management* is a strategy utilizing information gathered during the following processes:

• *Stakeholder identification:* Establishing a complete listing of affected parties either internal or external to program

- *Stakeholder analysis:* Recognizing stakeholder's needs, authority (influence, power), and interrelationships
- *Capturing expectations:* Recognizing the different set of interests and benefits each stakeholder expects to receive from the program

In my opinion, the most critical of the three processes is stakeholder identification. It has been my experience that often crucial stakeholders are not identified during program initiation, and this can lead to negative consequences. Take one of our recent K–12 (kindergarten and the first through the twelfth grade) programs as an example: the renovation of the Historic Stone School House at the Union Springs Central School District in Union Springs, New York. The immediate stakeholders that came to mind for us were the parents, teachers, school administration, and, of course, the students. Fortunately, as part of the initiation process, we spent considerable time developing a more complete listing of stakeholders, both internal and external to the school. The list was quite exhaustive, containing:

- Former, current, and incoming parents, teachers, school administrators, and students
- Local libraries
- Donors and charitable organizations
- Local taxpayers
- Local businesses
- State and local officials
- Local youth organizations
- School board members
- Teachers' union representatives
- The State Department of Education
- The State Historic Preservation Office
- The State Department of Environmental Protection
- Equipment and service vendors
- Elected officials
- Parent-Teacher Association
- Sports booster clubs
- Colleges and universities
- Local consortia or regional organizations
- Local contractors and design professionals

In New York State, the initiation of all K–12 capital programs is subject to a public referendum to secure bond financing. As a local resident, I can tell you that this project would not have been approved for construction by the voters without the key support of one important stakeholder, State Senator Michael F. Nozzolio, who secured $237,958 in state aid for the project. Thankfully, as part of the stakeholder identification process, we focused on him early as a key supporter.

Once a complete list of stakeholders is established, the next step is to determine which ones to focus on to help ensure success. There are two fundamental factors: the stakeholder's level of interest and the level of influence. As Table 1.4 illustrates, the team's energy should be focused on those individuals or entities that have a high level of interest and can most effectively influence the decision-making process. It has been my experience that politics plays a big role in public capital programs and invariably the important stakeholders will include public officials and politicians. These are the people with political interest and influence.

Once correctly focused, the program team must capture the stakeholder's expectations. There are several tools and techniques [6] to do this, but in my opinion, the most effective for capital programs is to establish a prioritized list of expectations for each stakeholder by directly discussing with them their issues, concerns, and, most important, the expected benefits of the program to them. It is important that the list is updated frequently because as the program goes through its life cycle (initiation-planning-execution-closure) issues and concerns may change, including the level of priority.

Once expectations are established for each important stakeholder, a communications plan is developed to inform and engage them on a regular basis. In addition, procedures for team building and conflict resolution are established to ensure stakeholder harmony. This is more of an art than a science and requires a program manager with a high level of emotional intelligence. Techniques such as effective listening, negotiating,

TABLE 1.4

Stakeholder Management

		Interest	
		High	**Low**
INFLUENCE	High	Focus on	Satisfy
	Low	Inform	Disregard

and influencing will need to be applied by the program manager to be most effective. The establishment of trust with each stakeholder is also paramount. This type of trust relationship is best defined as the willingness of one party to be vulnerable to the acts of another, not an easy thing to do. This is where the program manager must leverage his or her leadership versus management skills. More on that in Section 3.1, Chapter 3.

The last of the three themes of program management is benefit realization. The whole point of any program is to provide benefits to the stakeholders. Believe it or not, on some programs, this gets lost in the process. I believe that was the case on the Big Dig. For some of my recent capital programs, stakeholder benefits were:

- A better learning environment for the students at Union Springs Central School District
- Improved health of the community through cleaner drinking water by providing covered storage tanks to replace open reservoirs
- Increased recreational opportunities for the community through a cleaner Onondaga Lake in Syracuse, New York
- The reduction in Cornell's University's environmental footprint through implementation of LEED concepts on the new ecology building in Ithaca, New York
- An improved evacuation route during hurricanes for Brigantine Island and Atlantic City, New Jersey, through reconstruction and elevation of the highway system

I learned early in my career that, in marketing, features are not benefits. For example, it is better to tell a client that you will get the project done ahead of schedule and that will allow them to utilize the facility sooner and save money (benefit) than to tell them that you use the latest version of Primavera software for scheduling (feature). Benefits of a program are similar. From the examples above, it can be seen that a benefit is what the stakeholders have to gain from the successful execution of the program. As was mentioned earlier, a project delivers outputs (a new school) and a program delivers outcomes (a better learning environment).

Benefit realization encompasses three steps:

- Definition of each benefit
- Description of how benefits will be measured

- Establishment of the correlation between project outputs, program outcomes, and program benefits

If we go back to the Apollo program as an example and further study JFK's historic speech to the US Congress, we can see how brilliantly he outlined the most important benefit of going to the Moon: "If we are to win the battle that is now going on around the world between freedom and tyranny, the dramatic achievements in space which occurred in recent weeks should have made clear to us all, as did the Sputnik in 1957, the impact of this adventure on the minds of men everywhere, who are attempting to make a determination of which road they should take." The benefit of the Apollo program would be no less than freedom from tyranny for mankind! That definitely should have (and I believe it did) motivate the stakeholders (us).

After defining a benefit it must be measured to see to what degree it was actually achieved. A good way to illustrate how to measure a program benefit is through examples as well. For a K–12 school capital program it might be the improvement of student scores on mandated standardized tests. For the LIPO program the benefits were directly measured by sampling the water of Onondaga Lake to determine if oxygen levels were increasing and phosphorus levels were decreasing. For a program, it's not whether it was completed on time, within budget, and with a high level of quality (i.e., successful project attributes) but rather whether the intended outcomes and benefits were achieved. Did the program bring value to the stakeholders?

Sometimes it is difficult to establish a direct correlation among project outputs, program outcomes, and benefits. How does, for example, a successful K–12 renovation program improve student achievement? Communicating this correlation is important in getting stakeholder buy-in to the program. It must begin in the initiation phase during the determination of the project scope. A great way to do this is to use examples from previous programs. The US Green Building Council does a great job with this in its LEED for School campaign [10]. Based on research they can show that implementation of the LEED principles on K–12 capital programs in Washington State resulted in a 15% reduction in absenteeism and a 5% increase in student standardized test scores. So if improved student achievement is a proposed benefit, then LEED items, such as improved temperature control and high-performance lighting, probably should be priorities in the projects that comprise the program.

1.4.2 Program Management Life Cycle

Similar to a project, a program has a life cycle and set of process groups. The *Standard* defines the process groups as follows:

- *Preprogram Preparations:* Those processes performed to determine the need, feasibility, and justification for the program
- *Program Initiation:* Those processes performed to develop the program charter that outlines the desired outcomes and high-level roadmap for success
- *Program Setup:* Those processes performed to establish the management infrastructure including the schedule, budget, quality expectations, and the rules of engagement for the team
- *Delivery of Program Benefits:* The core work of the program through the execution of the component projects
- *Program Closure:* Those processes performed to ensure the controlled closure of the program

The program processes progress in a linear fashion with gate reviews at each step. Notice that the process groups for programs are similar to the process groups for projects. In my view, they are really equivalent things with different nomenclature as illustrated in Table 1.5. The difference is that program management processes address issues at a higher level and thus require different tools and techniques. The *Standard*'s nomenclature for the processes can be confusing, especially for construction professions where the three themes of program management and the triple constraints of project management interplay on a regular basis. For the sake of clarity and consistency, especially for my

TABLE 1.5

Project versus Program Characteristics

	Project	Program	
MONITORING AND CONTROL	Initiation	Preparation	
	Initiation	Initiation	
	Planning	Setup	GOVERNANCE
	Execution	Delivery of Benefits	
	Closure	Closure	

construction colleagues, I use the project processes nomenclature for programs as well.

1.5 CASE STUDY OF AN EFFECTIVE PROCESS-BASED MANAGEMENT APPROACH

Recently I had an opportunity to witness a process-based management approach work to perfection. The process was the management of air traffic control operations at LaGuardia Airport, the busiest crossed-runway airport in the world. Maybe even more impressive is the fact that the New York City metropolitan area's JFK International, Newark Liberty International, and LaGuardia airports combine to create the largest airport system in the United States and first in the world in terms of total flight operations. In 2011, LaGuardia handled just under 25 million passengers, JFK handled 47.4 million, and Newark handled 33.9 million, making for a total of approximately 105 million travelers.

I was very fortunate to be invited to witness the operations by LaGuardia's Federal Aviation Administration tower manager as part of a capital construction program we were doing at nine major airports across the United States. I had a bird's-eye view from LaGuardia's new 198-foot control tower of the operation that manages 1,200 flights a day.

The task of ensuring safe operations of all commercial and private aircraft falls on the air traffic controllers at LaGuardia. This team of five individuals must coordinate the movements of hundreds of aircraft a day, keep them a safe distance apart, direct them around bad weather, coordinate their activities with nearby JFK and Newark airports, and at all times ensure that traffic flows smoothly, both in the air and on the ground, with minimal delays. Wow! They make life-and-death decisions minute by minute for five to six hours a day, five to six days a week. So how do they successfully pull that off? Surprisingly through a process-based management approach that relies much more on human instinct and teamwork than the advanced computer and radar systems the profession has used for decades.

Believe it or not, at LaGuardia the primary method of doing this is visual observation from the control tower although they do have extensive radar and computer technology for confirmation. Judgment calls are made by the controllers regarding the separation of aircraft, such as the safe distance between the aircraft taking off, the one taxiing to the runway, and the one ready to land.

To make this work, the control tower crew is divided into three general operational disciplines:

- *Ground Controller:* Responsible for the airport "movement" areas, which includes all taxiways, inactive runways, holding areas, and transitional aprons or intersections where aircraft arrive, having vacated the runway or departure gate.
- *Local Controller:* Responsible for the active runway surfaces. The local controller clears aircraft for takeoff or landing, ensuring that prescribed runway separation will exist at all times.
- *Terminal Area Controller:* Responsible for the airspace surrounding the airport. Terminal area controllers ensure that aircraft are at an appropriate altitude and that aircraft arrive at a suitable rate for landing.

What fascinated me most about the process was the transition of responsibility between members of the tower crew as an aircraft entered the immediate airport environment. The crew records aircraft data and manages each flight on pieces of paper called flight progress strips. The strips are mounted in a plastic boot and physically passed (sometimes tossed) from one controller to the next as the aircraft enters his or her particular zone of responsibility. Analogous to a baton toss in a relay race (or a gate review in a program), this process ensures that there is no question which controller has taken responsibility for the safety of the aircraft at any point in its juncture from the air to the gate. They do these 1,200 times a day! You might think given this dynamic that the scene in the control tower is that of chaos. I can tell you that I experienced just the opposite; in fact the three-person crew and two-person management team work in complete harmony and what appeared to be a state of confident relaxation. They seemed to be more concerned with being a good host to me than the 747 approaching for landing with 500 people on board (see Figure 1.3).

I could not help think, "If a process-based management approach can work at LaGuardia for air traffic control operations, it ought to be effective on construction programs where there is much more time to plan and the consequences of failure are much less severe."

FIGURE 1.3
Air traffic control operations at LaGuardia Airport (2011).

1.6 CHAPTER SUMMARY AND KEY IDEAS

1.6.1 Chapter Summary

In this chapter the basic concepts of a process-based management approach were investigated and it was demonstrated how they were successfully implemented on a large and complex construction project as well as for air traffic control operations at one of the busiest airports in the world.

The contrast between strategic and tactical thinking was explored as was the importance of alignment between them. The basic definitions of project and program management were outlined. The *PMBOK* guide and the *Standard for Program Management* were outlined with a focus on how they can be successfully applied on construction programs.

1.6.2 Key Ideas

1. The PMI process-based management approach can be applied effectively to large and complex capital projects.
2. The PMI and the CMAA have competing definitions for construction program management. A far better way to describe the process is a definition that includes components of both. *Construction program management* is best described as "the practice of incorporating

both strategic and tactical management techniques, on one or more projects, from inception to completion, to increase the likelihood of success."

3. Program management requires both tactical and strategic thinking, and alignment between them. The difference between tactical and strategic thinking can be described as the difference between doing things right, and doing the right things.

4. Project management has been portrayed as the delicate balance between the triple constraints of scope, cost, and schedule and the ultimate level of quality. In a similar fashion, program management can be portrayed as the delicate balance between the three themes of stakeholder management, governance, and benefit realization, and the ultimate level of program value.

5. Stakeholder management requires processes to identify, analyze, and capture the expectations of any individual, group, or entity, who has something to gain or lose from the program. The team should focus on those who have both a high level of influence and a high level of interest.

6. Governance comes from the Latin word to steer, so it includes defining requirements and the setting of priorities. Proper program governance will also determine the program's justification, feasibility, and the readiness of the team.

7. Program benefits are items or effects that the stakeholders will gain from the successful execution of the program. It is the entire purpose of the program. Management of the benefit realization process entails three steps: the definition of each benefit, a description of how the benefits will be measured, and the establishment of the correlation between project outputs and program outcomes.

2

Program Management Process Groups

2.1 INTRODUCTION

In Chapter 1 the PMI's project and program management process groups were introduced with an emphasis on the contrast between tactical versus strategic thinking and the importance of alignment between them. In this chapter, we drill down into these concepts, and correlate them with the construction management (CM) processes that control the triple constraints of project management and balance the triple themes of program management. We also investigate the program management life cycle in more detail, specifically applied to construction.

2.2 PROCESS GROUPS AS CM PROCESSES

When I first started in the construction industry, on a plane trip to Kiewit's corporate office in Omaha, Nebraska for a T&D (training and development) session, I was fortunate to sit next to an industry veteran. I was full of energy and focused on my career ahead. He was reserved, thoughtful, and frankly what I considered as a bit jaded. He cautioned me that I was in for a very demanding career in a challenging profession. What stuck in my mind the most, though, was his portrayal of the construction business as synonymous with the Ford Motor Company developing high-performance vehicles, except each vehicle on the assembly line is unique with individual specifications and performance requirements. Looking back I think this was correct, and this applies even more so today, where LEED and other integrated design concepts are pushing us to develop facilities that perform to the ultimate level of energy efficiency and environmental control.

All construction projects and programs are inherently unique primarily because of environmental factors. These include the physical aspects of site location, the area's demographic–economic–political structure, the level and availability of required resources, and many other factors. Construction programs are also inherently challenging and complex. In fact you can argue, as the CMAA does, that construction projects are so complex and unique that they are actually programs.

I consider the complexity of managing construction related to the concept of entropy, the second law of thermodynamics. Historically, the concept of entropy evolved in order to explain why some processes occur spontaneously whereas their time reversals do not. In essence, it says that things do not easily become orderly. Have you ever seen a glass full of water fall off a table and break? I am sure you have, I am also sure you have not seen a broken glass jump up from the floor, reconstitute itself, and land back on the table full of water. Life does not work like that! Neither do projects or programs. The good news is that through proper planning, monitoring, and control, both projects and programs can be effectively executed. And for me, PMI's process-based management approach is the way to proceed.

As we explored in Chapter 1, all programs, regardless of type, have similar life cycles consisting of the process groups of initiation, planning, execution, and closure. Under PMI's approach, proactive measures (monitoring and control) are applied throughout to ensure adherence to the plan and to take corrective action if needed. All construction programs also have standard processes that directly align with these process groups. This, as is shown, will help us out of the entropy trap and bring a little order to the natural chaos of managing construction. I call this "putting your arms around the program." In this chapter we give each CM process an overview, and then drill down much farther in future chapters.

2.2.1 Initiation Process

In construction this is commonly referred to as "programming." In this phase the initial scope is established. This involves developing alternatives at the conceptual level, analyzing risk and opportunities, calculating economic payoff, identifying and analyzing stakeholders' interest and influence, developing a financial plan, deciding on the program organization and control, and ultimately making a decision to proceed (or not). In this phase it is critical to have both creative and technical input from the team as the various "what if" scenarios are vetted. This is also where the important work on the mission and vision statements takes place. These efforts culminate in the creation of the program charter. The program charter establishes and gets buy-in to the critical objectives and the rules of engagement. This is also where the mission and vision statements are fine-tuned with input from the entire team. A communications process is developed and implemented. Responsibilities are assigned and authority established. The program charter is developed in a collaborative way and results in a formal document signed by each member of the team.

Unfortunately, it is still common practice for owners today to perform the tasks required for the program charter internally without assistance from their design or construction professionals. This can lead to negative consequences, as the old adage, "Garbage in, garbage out," applies. In other words, decisions regarding the program scope without input from the experts may be based on false information and this, of course, cannot be positive.

On one of my recent K–12 programs that is just what happened. In that market, the initiation process is referred to as the prereferendum phase, as it is the time the school district plans for the project prior to putting it out as a public referendum for approval. In this case the school district identified, in the referendum, priority scope items valued at $11.5 million. Had it engaged a CM or even a cost consultant early on, it probably would have known the actual bid cost for these scope items would be $13.6 million. I am sure it was no fun for the District Superintendent and the Board of Education to inform the public that priority items they voted on, and were promised in the referendum, could not be included in the program because the preliminary budget was incorrect.

2.2.2 Planning Process

In construction, planning typically begins when the owner engages the services of a design and construction professional to develop the initial scope fully. Once a decision has been made to proceed with the program, the planning phase becomes the most critical phase in the construction management life cycle. Issues that are not addressed in this phase will become much more difficult to overcome in future phases. This is the reason that the planning phase is often called the most critical in the program life cycle. There is an example of the negative consequences of poor planning later in the chapter.

For construction programs the key steps in the planning process are the following:

- *Creation of the Master Schedule.* The master schedule establishes the overall duration for the program and the critical intermediate milestones and deadlines. The best approach is through creation of a CPM schedule that details the entire timeline of the program and each project, from initiation to closure. This will include preconstruction items such as permitting, environmental reviews, and other governmental approvals, a detailed fragnet for passage through the design gates (conceptual design–design development–construction documents), a procurement schedule for both program- and project-level activities, a conceptual construction schedule for each project, and critical postconstruction tasks such as substantial completion and formal acceptance. The master schedule will also include the prerequisites to start and close the component projects as well as the interrelationships between them. Very critical in this process is the development of a conceptual construction schedule for each project. Unfortunately (I will have to admit that this stage often reveals a pet peeve of mine) many preliminary master schedules have a detailed design fragnet but contain only one task called "construction," outlining the entire execution phase for the project. In construction, setting of the project duration and critical intermediate milestones is extremely important for proper planning, monitoring, and control. Without a conceptual construction schedule how can you do that?
- *Creation of the Master Budget.* Creation of the master budget involves the estimation of soft and direct costs, and the establishment of program and project level contingencies. "Soft costs" is a construction term for those budget items not considered direct construction costs such as architectural, engineering, and legal fees, financing, and other pre- and postconstruction expenses. Direct costs are those budget items that can be directly attributed to the physical construction work. In construction, the majority of program-level expenses fall into the soft cost category, whereas project-level expenses average between 20%–25% for soft costs. Contingencies are allowances

established at each phase of the construction management life cycle to account for unknowns. Considering the concept, "time is money," none of these costs can be properly budgeted until the master schedule is fully developed and an overall duration for the program is established. This should be obvious, as you would agree that a two-year program will have a different cost than the same program scope completed over three years. Unfortunately, my experience has been that preliminary master budgets are often crafted without any consideration of the schedule. More on that later.

- *Establishment of Quality Standards.* For construction, the level of quality is defined in terms of each project's physical quality (workmanship, level of finish, aesthetics, etc.), fulfillment of management objectives (cost, schedule, scope, impact on operations, etc.), functionality (including life-cycle considerations), and the alignment of each project with the program's expected outcomes and benefits. Because of the interplay between the triple constraints of project management and the three themes of program management, in effect, the quality standard sets the baseline for all other program aspects. Earlier when I was a contractor, we used to tell our clients that if they wanted a higher standard of quality (i.e., a Cadillac instead of a Chevrolet), something would have to give: either the cost would have to go up, the scope would have to decrease, or we would need more time. Of course they did not like to hear that as they thought they had already ordered a Cadillac.

- *Establishment of Safety Standards.* Construction is a dangerous business. There are a number of hazards and risks that site workers are exposed to on a daily basis. Those working in the vicinity of heavy machinery risk being crushed by excavator arms or run over. Electrocution is a risk for those working near live cabling that could result in severe injury or death. We all have heard about the risks associated with asbestos, mold, and lead paint exposure. So it is not surprising that according to the National Institute for Occupational Safety and Health, 8% of the US workforce is in the construction industry yet they account for more than 22% of all industry fatalities. Not a good ratio! Because project risks associated with site safety are so significant in construction, focus on setting safety standards during the planning process is critically important. The Occupational Safety and Health Act (OSHA) was passed in 1970 to protect employees by stating that employers have a legal obligation to provide their employees with a safe working environment. The OSHA has published legal requirements that define how employers must protect their staff, entitled, "Safety and Health Regulations for Construction." Safety may be the only program planning task that is legally required.

- *Determination of the Project Delivery Method.* The proper planning of a construction program involves many key, early decisions, but none more important than choosing the best project delivery method. Because the project delivery method sets, among other things, the rules of engagement for defining, monitoring, and controlling accountability within the construction team, a poor choice early on can doom a construction program. The selection of the best delivery option includes more options today as the emergence of alternative project delivery methods, such as design–build and CM-at-risk, claim to address the weaknesses in the "traditional" design–bid–build model.

The design–bid–build approach is commonly referred to as the "traditional model." This is a bit of a misnomer though, as it is relatively new. It actually came about from the construction of the Transcontinental Railroad, which suffered from unscrupulous and illegal acts among the designer, contractor, construction management firm, financing company, and federal government. One consequence of the scandal was the formal separation of design services from construction work on federal projects through an act of Congress in 1893, and, ultimately, legislation at both the federal and state levels requiring use of the design–bid–build approach where public money is used. The design–bid–build approach establishes a linear, and prerequisite, relationship between the discrete project phases. Separate entities perform design services and construction work and design is required to be completed prior to the start of construction. Commonly, the designer—or an agency construction manager—oversees the work of the contractor to ensure quality and being on time, on budget completion. By clearly separating the roles and responsibilities of the team, the design–bid–build approach sets checks and balances and enhances the accountability of the contractor and designer toward the owner. Commonly, under this scenario, design services are awarded based on a qualifications-based selection process, and construction work is awarded based on the lowest responsive, responsible bid.

The design–build model's origin can be traced to the master builders of the great pyramids some 7,000 years ago. Under this approach, a single entity is responsible for designing and building the construction project, similar to other modern industries such as aircraft and automobile manufacturing. Currently, design–build is the fastest growing project delivery method, and according to a recent study [11], represented over 40% of nonresidential construction projects in 2011. This is an impressive percentage, given the fact that it was not until 1996 that legal authority was granted to federal agencies to use the design–build method. The two major advantages of the design–build model are speed of construction and a single source of responsibility. The primary benefit is facilitating fast tracking of a project. Fast tracking is often the most effective way to shorten the duration of a project, by allowing activities originally scheduled in sequence, such as design and construction, to overlap. And because time is money, fast tracking can result in significant cost savings as well. The primary disadvantage of the single source of responsibility is the loss of checks and balances and, ironically, an unintended negative impact on accountability. Owners sometimes cite a lack of control during design, which can result in reduced quality and can affect the ultimate scope of the project. Design–build services are commonly procured based on a combination of a quality-based selection process and various negotiated "best value" scenarios. As such, the design–build approach is sometimes criticized for public projects because it does not permit competitive bidding of completed plans.

CM-at-risk has been used in the private sector for many years. In fact, in New York City it is the primary delivery method for commercial projects. Over the last 10 years, states such as New York, Florida, Texas, California, and Arizona have pushed for and enacted laws allowing the use of CM-at-risk in the public sector. Similar to the traditional method, separate entities perform design services and construction work but unlike the traditional method, design is not required to be completed prior to the start of construction. This is because the CM-at-risk firm comes on board before the design and bidding documents are completed and works collaboratively with the design team to develop the project to a point where a guaranteed maximum price (GMP) can be developed. Fast tracking is possible because discrete portions of the design can be expedited, thus critical elements are constructed first. Commonly, the CM-at-risk is procured through a quality-based selection process. Once the GMP is established (anywhere from 30%–90% of final design), the CM-at-risk takes on the role of a general contractor, guaranteeing the price, quality, and schedule for the work. Advantages attributed to this approach include the engagement of a construction professional during design and the ability to obtain a GMP early in the process. As with design–build, however, CM-at-risk is sometimes criticized on public projects because it does not permit competitive bidding of completed plans, and it may be difficult for the owner to evaluate the GMP and determine if the best price has been achieved.

There are a number of additional alternative project delivery methods, including design–build–operate–maintain, build–operate–transfer, and integrated. A brief description of each is provided in Table 2.1. Project delivery methods have been developed in an effort to find a better way to ensure the successful execution of construction projects. Regardless of the method used, the proper execution of project management techniques and principles is required. In addition to accountability, the owner's risk tolerance and level of expertise, as well as regulatory or funding requirements, may also play a consideration in the method chosen.

- *Selection of the Document Control System.* In 1597, during the scientific revolution, Sir Francis Bacon coined the phrase, "Knowledge is power." The basic concept is that when you are in possession of knowledge that others do not possess you are at an advantage. Today's web-based construction document control systems provide that

TABLE 2.1

Additional Project Delivery Methods

Project Delivery Method	Description
Design–Build–Operate–Maintain	This takes the turnkey approach of the design–build model one step further by including the operations and maintenance of the completed project in the same original contract.
Build–Operate–Transfer	Under this approach, the same contract governs the design, construction, operations, maintenance, and financing of the project. After some concessionary period, the facility is transferred back to the owner.
Integrated Project Delivery	Integrated project delivery is an emerging business model that allows for the entire construction team (owner, contractor, and architect) to collaborate during initiation to make the most effective decisions. Under this project delivery method, behavior and contractual relationships are enhanced by leveraging building information modeling (BIM). BIM helps project teams use consistent, reliable information in a common collaborative environment, increasing understanding of design intent and improving efficiency.

advantage. These systems are a clearinghouse for all construction documents, commonly referred to as the project record. The project record includes daily reports, submittals, requests for information, meeting minutes, contracts, pay requests, change orders, and literally any document generated by, or related to, the program or the component projects.

The power of these modern communication tools is not only the ability to "slice and dice" the data and communicate them at the speed of light, but also to include key accountability features such as *ball-in-court* and *issue tracking*. In construction, documents are approved, changed, updated, and reapproved all the time, often being sent back and forth between multiple entities. The ball-in-court feature keeps track of where a document is, and who is responsible for moving it along. The benefit is faster turnaround times. This is especially important on product and material submittals, because on many programs the procurement process is on the master schedule's critical path.

The issue feature is essentially a "flag" placed on those construction documents that relate to an important issue. Issue tracking is especially important when unforeseen conditions happen, as the documents associated with such changes often follow a predictable paper trail. An example from a previous project of mine was the discovery of an underground petroleum storage tank in the way of an added footing. The paper trail included documentation of it in a daily report, a request-for-information on how to proceed, meeting minutes of the decision-making process, and finally a directive of how to proceed. The program benefits from having all this information at the touch of a finger's expedition of the decision-making process. Had this relatively minor issue not been expedited it could have affected or delayed the project schedule. That, again because of the concept of "time is money," would have been much more costly to the project then the remediation effort itself.

It is noteworthy that there are many excellent web-based document control systems today specifically tailored to the construction industry. It is important to coordinate the selection of the system with the communications plan developed in the initiation phase.

- *Development of the General Requirements.* The general requirements, also referred to as the "front-end documents," "general conditions," or "boiler plate," are the formal rules of engagement between the owner and construction team. The general requirements are used for scope items that do not apply directly to the physical construction work. They include provisions for items such as overhead and profit rates, construction time requirements, insurance requirements, trade wage requirements, bond requirements, permit(s) responsibility, procedures for change orders, temporary utility responsibility, safety responsibility, scheduling requirements, and dispute resolution procedures. There are several organizations that produce standard general requirement documents for the various project delivery methods. These include the CMAA, the Design Build Institute of America, the American Institute of Architects, and the Engineers Joint Contract Documents Committee (EJCDC). I prefer use of the EJCDC general requirements as they were developed through a collaboration effort among the American Council of Engineering Companies, the National Society of Professional Engineers®, the American Society of Civil Engineers, and the Associated General Contractors of America. I like collaboration in general, as I believe everyone in construction, regardless of his or her role, should be treated fairly and equally.

There are two important things to keep in mind regarding these standards. First is that they are not meant to be used verbatim. Revisions to the standard documents are required to fit the requirements of the specific program and component projects. Second, alignment of the provisions with the core values of the team is critical. Which is more important to the team, an environment of teamwork or one of accountability, or are they equally important? Potentially controversial items or divisive issues such as liquidated damages, the change-order process, and alternative dispute resolution procedures, need to be properly vetted by the entire team. The consequences of not doing this during the planning process can disrupt team harmony and negatively influence the progression of construction during the execution phase.

- *Development of the Technical Specifications and Plans.* A colleague, Bill Sands, a principal at Bearsch Compeau Knudson Architects & Engineers, PC in Binghamton, New York, often presents with me at my all-day seminars on construction project management. In his part of the presentation, Bill likens the design phase of a construction project to a challenging and complex project in and of itself. By the classical definition of projects (i.e., they are temporary and produce a unique product, he is absolutely correct. In fact, the design phase, at least from a scheduling perspective, is entirely deliverable with the final result being the completed technical specifications and plans (construction documents).

Regardless of the project delivery method, the design phase will include three major deliverables: conceptual design, the design development drawings, and the construction documents. The conceptual design presents the preliminary scope in a form the owner can understand and accept. It will illustrate the scale and relationships between program and project components. The design development documents refine the conceptual design and fully describe all aspects of the scope so that all major issues are resolved that could cause redesign during the construction document phase. The design development documents are commonly computer-drafted (using the latest AutoCAD Architecture or Revit BIM Architecture software) to scale drawings that illustrate the project as it will look when constructed. Outline technical specifications are developed that will begin to delineate specific acceptance criteria or performance requirements for materials or equipment. The construction documents are the final set of technical specifications and plans suitable for obtaining building permits and competitive bidding or pricing.

Each of these deliverables is often placed as a critical intermediate milestone in the master schedule and is referred to as a "design gate." They get this name from systems engineering where a "decision gate" is described as a formal way to conclude and accept a deliverable before moving on to the next one. Remarkably, in my experience, the percentage of the total design effort for each deliverable is consistent across almost any type of construction work. As a rule of thumb, conceptual design is about 20% of the overall effort, design development is 30%, and construction documents are 50%.

At each of the design gates, the budget, schedule, and quality of design are re-examined to ensure conformance with the program objectives. For the budget and schedule constraints, detailed estimates and plans are done, and corrective action is taken if there is a variance from the baselines. For these constraints, corrective action generally will take the form of value engineering, constructability analysis, and, as a last resort, reduction of scope starting with the low priority items.

Value engineering is best described as finding a less expensive way to achieve the same thing, again without jeopardizing quality. A good example from one of my recent programs was the standardization of embedded polyvinylchloride (PVC) conduit as opposed to the specified embedded steel conduit for the concrete slabs on grade. This saved on both material and labor cost, and saved schedule time as the PVC conduit could be easily flexed into place around reinforcing steel and embeds, whereas the more expensive steel conduit required the labor-intensive work of a pipe bender.

Constructability analysis is best described as finding an easier way to do the same thing without jeopardizing quality. A good example from one of my recent projects was the substitution of precast concrete planking for the specified cast-in-place concrete for the ceiling of a mechanical piping gallery. The use of the planking allowed for preassembly and installation of large sections of the mechanical piping prior to construction of the concrete ceiling system. This saved cost, schedule time, and improved quality as it provided for better working conditions for the trades.

- *Development of the Procurement Documents.* Vendors are employed in construction as a means to transfer or share risks. Vendors typically include, at a minimum, a design entity who takes responsibility for the design, a construction professional who takes responsibility for the performance of the program, and a trade contractor who takes responsibility for various aspects of the construction work itself. The project delivery method will determine the structure of agreements between the entities and the procurement strategy (competitive bidding, qualifications-based selection, etc.) will determine the nature of the relationship between the vendors and owner. If we over-simplify for the sake of brevity, there are really just three major things to consider when deciding on a construction vendor: how will the vendors be solicited and chosen, how will project risk be transferred, and how will the chosen vendor be treated. In construction there are standard processes to do just that. The processes involve three components: the bidding documents, the contract documents, and the construction documents. Unfortunately, I have noticed that even seasoned construction professionals interchange these terms. Regardless of the project delivery method or the procurement strategy there is a standard process (and nomenclature!) for selecting vendors for construction programs.

Development of the procurement documents starts with the bidding documents, which are the most comprehensive, and are used to solicit and record bids from vendors, but they also include the contract agreement that will be executed by the selected bidder. The bidding documents include items such as the invitation to bid, instructions to bidders, information to bidders, bid forms, and, if required, a bid security (insurance against bidder default).

The contract documents include the contract agreement and, if required, the performance and payment bonds (insurance against contract default).

The construction documents include the general requirements and technical plans and specifications, which will be developed to various degrees of completion based on the vendor, the project delivery method, and the procurement strategy.

I would be remiss not to note, especially for publicly funded programs, the issue of legal accountability of the owner when selecting construction vendors. Federal antitrust legislation, including the Sherman Act of 1890, the Clanton Act of 1914, the Federal Trade Commission Act of 1914, and the Robinson–Patman Act of 1936, established strict rules for soliciting and selecting construction vendors. Essentially these rules have to deal with price-fixing and bid-rigging. It is the responsibility of the owner to ensure these requirements are being followed and included in the bid documents.

2.2.3 Execution Process

During my training sessions I always get a laugh from the audience when I describe the construction phase of a capital program as the "execution phase." My construction colleagues will confess that sometimes it does feel just like an execution! The execution phase in construction is where the "tires hit the road" and the physical work takes place. As discussed earlier, it will feel much less like an execution if the planning phase is properly managed, and the majority of potential issues and pitfalls are addressed then.

Unfortunately, I have a great example of where a critical issue during the planning phase was not properly addressed and almost completely derailed us during construction. It happened on the reconstruction of the 25 bridges that comprise the I-84/Route 8 interchange in Waterbury, Connecticut. It was 1989 and my very first construction assignment with Kiewit. The bridge program included $17 million for sandblasting and painting each bridge structure. I remember how eager I was to start my career with a "meaty" assignment and that I actually lobbied to manage this work. Sometimes you have to watch what you wish for!

In 1989, the standards for lead abatement on bridge structures were evolving in Connecticut and nationwide. The project specifications included language that stated, "a minimum of 75% of the sandblasting debris must be collected, removed from the site, and properly disposed of in accordance with state and federal regulations." The problem was that the sandblasting debris contained very high concentrations of lead and the interchange was literally located on top of the Naugatuck River. Long story short, we were directed to collect 100% of the debris, and this resulted in a severe financial burden for our two painting subcontractors, one of which was a woman-owned business enterprise, who eventually defaulted on her contract. This led to a lawsuit that kept us in the courts for 10 years, well beyond my tenure with Kiewit. What a way to start a career!

Looking back as a "Monday morning quarterback," had the planning phase been properly managed, this could have been completely avoided. Think of the vagueness in just the one line of the painting specification noted above. How do you measure a minimum of 75% of sandblasting debris? Is it fine for a maximum of 25% of the sandblasting debris to enter the Naugatuck River? What if the Connecticut standards for handling of lead paint debris differ from the federal standards (they did by the way)?

In construction programs the key steps in the execution process are:

- *Approval for the Commencement of Projects or Components.* In this phase it is determined whether the prerequisites established in the planning process have been met in order for either a component (i.e., conceptual design) or component project (i.e., the first bridge renovation) to begin. It also involves assessing the readiness of the team and assigning responsibility for the discrete activities.
- *Managing of Program Outcomes and Project Outputs.* In this phase it is all about ensuring, as things are being implemented, that the expected program outcomes and the component project outputs are being met. This is done in a proactive way by periodically reviewing the overall program status to ensure benefit realization is on track and that each component or component project is in alignment with the objectives. The project team does this by ensuring that the proper resources are available and the performance of the team is regularly assessed.

For complex construction programs it is also important that the interfaces between the component projects are properly managed. The hierarchical structure for component projects is referred to as the program work breakdown structure (PWBS) and the correlation between the component projects is referred to as the program architecture. The PWBS and the program architecture together outline the characteristics, capabilities, deliverables, and timing for each component project as well as the critical interfaces between them. An example of such an interface from the LIPO program was the relationship between the construction of the secondary effluent pumping station (the SEPS project) and start-up of the Metro project. Without the SEPS, the Metro project could not receive the required flow of effluent for processing. And without the Metro project online, the LIPO program would not succeed in cleaning up Onondaga Lake. In addition to physical links such as the example above, interfaces may also include items such as sharing of resources; the result of poor, or exceptional, performance on one project on another; and the social items such as the impact of project team morale on program team morale.

• *Monitor and Control Changes.* One central theme of John Steinbeck's *Of Mice and Men* is that even with the best of plans things happen unexpectedly. This is certainly the case in construction. Because we know things are going to change, we must, ironically, plan for the unknown. Planning for change requires three basic steps: setting a baseline, setting procedures to monitor the baseline, and establishing a process for taking corrective action when there is a variance for the baseline. For construction projects there are several standard change processes that have been developed and implemented successfully. During the planning phase these will be incorporated into the procurement plan and will be an important part of the bidding documents.

The experienced construction professional knows that procedures for changes external to the projects, and more likely associated with the program, must also be planned for and managed. Examples of such changes include shifts in macroeconomic conditions (i.e., the Great Recession), world events (i.e., 9/11), and politics (i.e., Republican vs. Democratic rule in the United States). Such changes may not just affect the execution of the program but may also determine its continued feasibility. It is hoped that, during program initiation, such changes were vetted during risk assessment and plans were developed to handle them.

2.2.4 Closure Process

My cousin runs a site-work and landscaping business that I have used on a few of my construction projects over the years. I feel bad for him because, inevitably, the scope of his landscaping work is cut back as it is one of the last opportunities to cut costs and schedule time. The closure process is similar in nature. Because of its position in the program life cycle, it seldom gets the attention that is required to do it fully and correctly. Properly done, the program closure will consist of the following processes:

- *Final Program Reports.* These reports are an accounting of success and failure and lessons learned, and include an archive of the program record. The reports themselves are useful as learning tools for others through knowledge transition and can become critical documentation for claims. However, I believe the real value is gained by the project team in the work required to prepare the reports. Formalizing lessons learned in writing (as in a book) is one of the best ways to develop as a professional.
- *Formal Program Closure.* Successful construction programs are formally closed by completing each component project successfully and fully delivering on the program benefits. Standard procedures for the component project closures have been well developed for the construction industry, and include the critical intermediate milestones of beneficial occupancy, substantial completion, and final acceptance. Detailed procedures and rules for establishment of the criteria requisite for each milestone have also been developed. This is not the case for program closure and benefit realization. Whether the program achieved its intended value is often subjective. This can be problematic, as program closure requires the release of resources and the transition to operational status. More on that later in future chapters.

2.3 CONSTRUCTION PROGRAM MANAGEMENT LIFE CYCLE

Although each program phase is a prerequisite to the next, as a process-based approach requires, the program life cycle is not linear in nature. As the program progresses through the CM phases and things inevitably change, corrective action may require the team to revisit decisions made in a previous phase. I liken this to the "control–alternate–delete" scenario. It is important, and not necessarily common sense, that once the CM life cycle is "rebooted" that the phases proceed in a linear fashion again. Sometimes during the pressure and momentum of the execution phase, it is tempting not to follow this requirement fully. I think the Big Dig in Boston, Massachusetts might be such an example. An example where a reassessment of a program was properly made during the execution phase was the Superconducting Super Collider (SSC). Also nicknamed the Desertron [12], the SSC was a particle accelerator complex under construction in the vicinity of Waxahachie, Texas that was set to be the world's largest and most energetic. The program was cancelled in 1993 after the US Congress had spent $2 billion on planning and construction. Construction was actually going very well, as I remember my colleagues at Kiewit had set four world tunneling production records at the site. There were several reasons for the cancellation but the most significant was the loss of the proposed benefit: the need to prove the supremacy of American science with the collapse of the Soviet Union.

I created Figure 2.1 to illustrate this "control–alternate–delete" concept. The important concept to remember is that when things change, the team must revisit the decision-making process, sometimes all the way back to the program initiation phase!

2.4 CHAPTER SUMMARY AND KEY IDEAS

2.4.1 Chapter Summary

In this chapter the program management process groups were outlined as CM processes and it was demonstrated how they function in the life cycle of a construction program. Future chapters further detail each CM process and focus on the state-of-the-art tools and techniques used to monitor and control them effectively.

2.4.2 Key Ideas

1. The complexity of managing construction is similar to the concept of entropy. Without proper planning for monitoring and controlling "chaos," a construction program cannot be successful. The PMI

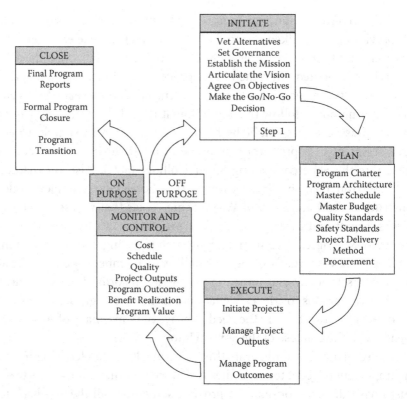

FIGURE 2.1
Program management life cycle.

process groups (initiation–planning–execution–monitoring and control–closure) can be directly correlated to standard construction management processes. This provides the structure needed to turn the chaos into manageable program elements.

2. The CM initiation process is commonly referred to as programming and is the phase where the initial scope is developed. It involves developing alternatives, identifying and analyzing stakeholders' interest and influence, developing a financial plan, deciding on program organization and control, and ultimately making a decision to proceed (or not).

3. The CM planning process typically begins when the project sponsor engages the services of a design and construction professional to develop the initial scope fully. Key steps in the planning process include creation of the master schedule, creation of the master budget, establishment of quality standards, establishment of safety

standards, determination of the rules of engagement including the project delivery method(s), selection of the document control system, and the development of the procurement documents including the general requirements and technical specifications and plans.

4. During the construction execution phase the "tires hit the road" and the physical work begins. This is where the majority of monitoring and controlling takes place to manage changes to the program plan as they occur. Planning for change requires three basic steps: setting a baseline, setting procedures to monitor the baseline, and establishing a process for taking corrective action when there is a variance from the baseline. During execution these processes are implemented to keep the program on track to realize its anticipated outcomes and benefits.

5. The CM closure process is a set of procedures for properly and formally ending the program. These include documenting success and failures, lessons learned, and creating the program records archive. Successful construction programs are formally closed by completing each component project successfully and fully delivering on the program benefits.

3

Initiation Process

3.1 INTRODUCTION

Construction programs generally start as a result of one of four things: a need, a problem, a mandate, or an opportunity. A need might result in the construction of a new airport terminal, like the $1.3 billion Terminal Expansion Program in San Jose, California, designed to alleviate congested air and ground traffic (and also improve Silicon Valley's world image). A problem of a dirty lake might result in a mandate to upgrade a county's wastewater treatment facilities, as with the LIPO. Or an opportunity might result in the construction of a microchip manufacturing campus, such as the $6.9 billion Globalfoundries Program underway in Albany, New York.

Regardless of motivation, construction programs demand an incredible amount of resources, entail significant risks, and require the collaborative skills and efforts of a diverse group of individuals and entities to be successful. And things can go horribly wrong without the right focus and direction.

The initiation process is where the collective ideas of the team are channeled and a path is chosen to achieve success. As one of my colleagues put it, "It's like herding cats." A successful program initiation process requires both strategic and tactical thinking and the unique skills of a leader.

3.2 MANAGEMENT VERSUS LEADERSHIP

Warren Bennis, an author and leadership expert, makes a distinction between a manager and a leader this way, "The manager asks how and when; the leader asks what and why." For me it's all about the difference between strategic and tactical thinking. In construction we have a lot of great tactical thinkers but very few true strategic planners. Not sure why that is, but it is absolutely the case. I think this is the main reason for the industry's poor program success rate.

Strategic planning is critically important during the initiation phase of a construction program as it will determine the justification and feasibility of the endeavor. If we look at the Apollo program again as an example, not only can it take a great deal of insight but sometimes require courage as well. How many would have thought it was feasible to put a man on the moon back in 1964? Even today, Apollo's feat would be extremely ambitious. And back then the onboard computer had less computing power than a modern digital watch [13]. They did it through great leadership.

Leadership has been described as a journey of discovery [14]. An effective program initiation process should feel the same way. In construction, the program initiation process demands the harnessing of the collective energy of the entire team. This "herding of the cats," so to speak, requires a program manager with a unique attribute, emotional intelligence. It was Daniel Goleman who first brought the term "emotional intelligence" to a wide audience [15]. In his research he found that in business, although qualities that are traditionally associated with great managers, such as intelligence, toughness, and determination are required for success, they are insufficient. Great managers must also be leaders who are distinguished by a high degree of emotional intelligence, which includes self-awareness, self-regulation, motivation, empathy, and social skill. This is true in construction program management as well.

The basic skill set for project managers is centered on order and control and makes use of standardized processes. The skill set for program managers is this as well, but it also includes leveraging the power of influence or persuasion [16], and when required, demanding what is needed. For construction programs it is about commanding attention, changing minds, and persuading decision makers and stakeholders to pursue the right path. This is critically important during early program choices, such as go/no-go decisions, because a mistake at this point may be irreversible.

In his second book *Primal Leadership*, Daniel Goleman, describes six different styles of leadership:

- *Visionary.* Visionary leaders articulate where a group is going, but not how it will get there. This sets people free to innovate, experiment, and take calculated risks.

- *Coaching.* Coaching works best with team members who show initiative and want more professional development. But it can backfire if it's perceived as "micromanaging" and can undermine self-confidence.
- *Affirmative.* This style emphasizes the importance of teamwork and creates harmony in the team by connecting people to each other. Goleman argues this approach is particularly valuable "when trying to heighten team harmony, increase morale, improve communication or repair broken trust in an organization." But he warns against using it alone, because its emphasis on group praise can allow poor performance to go uncorrected. "Employees may perceive," he writes, "that mediocrity is tolerated."
- *Democratic.* This style draws on people's knowledge and skills and creates a group commitment to the resulting goals. This consensus-building approach can be disastrous in times of crisis, when urgent events demand quick decisions.
- *Pacesetting.* In this style, the leader sets high standards for performance. The leader is obsessive about doing things better and faster, and asks the same of the rest of the team.
- *Commanding.* This is the classic model of "military"-style leadership: it rarely involves praise and frequently employs criticism. It is most effective when a sense of urgency is required.

In my opinion, the most effective construction program managers have a commanding leadership style, but can move among the approaches if necessary, adopting the one that meets the needs of the moment. A commanding, or even aggressive, leadership style is often necessary because in construction, it takes power (or at least perceived power) to get things done. It's not as bad as the old days, when the loudest guy often won, however, without power, you can be viewed as impotent, irrespective of your talents or the righteousness of your decisions. And that can lead to unfortunate results. I do not take this as far as one of my clients did recently when I was asked to meet with him to discuss an "issue" he had with one of my guys. For context, the client had been a successful developer and was now the director of a $140 million construction program. His issue was, "Your guy is always looking for win–win situations." I scratched my head and thought, "Boy, the world changes slowly!"

Developing trust with the program team is also paramount. My favorite definition of trust is "the willingness of one party to be vulnerable to the acts of another." No easy task to pull off, especially in the alpha-male-dominated construction world, but absolutely necessary to lead the team. Thankfully there are some techniques that can help develop trust. Skills such as effective listening (realizing what the other person is saying is more important than your response), empathy (understanding, and entering into, the feelings of another), and leading by example are a few critical ones.

During the program management life cycle (initiate–plan–execute–close), being an effective leader is never more essential than when navigating the choppy waters of the program initiation process. This is true because the range of possibilities is never greater, and the risk of taking the wrong path is never more fraught with potential consequence.

3.3 DEFINING THE ULTIMATE PURPOSE

A construction program must be designed, planned, and properly executed to achieve its desired benefits. But before that can happen, the ultimate purpose for the endeavor must be agreed upon, and committed to, by the team. The term PMI uses for this is the *program mandate*, which describes the strategic objectives and benefits the program is expected to deliver. Construction programs are initiated by defining this purpose in conjunction with establishing a valid business case. It's critical that the business case be in alignment with the program mandate. An example of the negative consequences of the business case "out of whack" with the program mandate is the K–12 capital program in New York State.

New York State is made up of 742 school districts (districts) that are individually responsible to their constituents to provide a "sound basic education" as mandated by the state constitution. All capital programs are required to meet the purpose of this mandate. So the ultimate purpose is clear. Districts fund their operations through two main sources. Operating costs are funded through tax levies based on a publicly approved yearly school budget. When a district decides to undertake a major construction program they generally will utilize multiyear financing, which requires a separate referendum approved by its taxpayers. Because cost is so important to voters, the construction program budget often gets determined more by what can be approved rather than an amount sufficient to meet the educational goals of the district. So the driving purpose for the program (providing a "sound basic education") often gets lost in the democratic process of mining votes, and as a result the needs of the children suffer. One way to ensure this does not happen is through skillful governance. Governance consists of three major components:

- *Defining the mission.* What is the ultimate purpose?
- *Setting the vision.* What would success look like?
- *Developing the strategy.* How to get there.

An example of skillful governance is the recently completed $48 million DCMO (Delaware–Chenango–Madison–Otsego) BOCES program. The term BOCES stands for Board of Cooperative Educational Services. There are 37 BOCES in New York State providing shared programs that serve children from their surrounding K–12 school (component) districts. By focusing on collaboration and cost containment, the BOCES help to relieve the financial burdens placed on local taxpayers. In addition to this important service, the BOCES also provide learning centers focused on vocational education and hands-on training.

The DCMO BOCES program included the complete renovation of, and large additions to, two learning centers in the Southern Tier region of New York State (see Figure 3.1). The program budget was split between two campuses located 35 miles apart. Each of the 17 component districts that comprise this BOCES were required to approve the program by presenting a bond referendum to their local taxpayers. No easy task for one district to agree on a capital program, so imagine the difficulty of getting 17 districts to come to a unanimous agreement!

This required leadership in developing and getting buy-in to compelling mission and vision statements for the program. The DCMO BOCES program mission and vision statements were:

- *Mission Statement.* Enhancing the quality of education through shared services.
- *Vision Statement.* The new facility should look and feel more like the student's future work environment than a school. In addition to technical skills, the facilities must be designed to encourage development of SCANS skills (these are broad academic and workplace skills developed by the Secretary's Commission on Achieving Necessary Skills). State-of-the-art, not cutting edge, technology, will be utilized. Aspects of LEED, including enhanced learning environments

FIGURE 3.1
DCMO BOCES' Harold campus.

and energy efficiency are to be incorporated into the design. It is to be built on time, within budget, and to a high level of quality. Enrollment is up so it is essential that the new facilities come online as soon as possible.

This is an excellent mission statement as it is "short and sweet" and is focused on long-term goals. A great mission statement should be easily remembered, readily agreed upon, and inspire. This program also had an outstanding vision statement as it was focused on specific benefits and outcomes. I also like the use of the words "look and feel" as they describe the experience of success.

As far as the third leg of governance, "strategy," the DCMO BOCES program team devoted over 3,000 person-hours developing the programming documents. The completed document can be seen in Figure 3.2. We explore the process for developing effective program strategies in detail in Chapter 4. The program was ultimately completed on time and within budget, and today the two campuses are considered some of the best of the 37 BOCES facilities in New York State.

FIGURE 3.2
DCMO BOCES' program documents.

3.4 DETERMINING JUSTIFICATION AND FEASIBILITY

The mission statement for our family business is, "We will change the world and make it a better place." It is easy to argue that it is justified, but is it feasible? It is critical on a construction program, even with the best of intentions, that it is determined to be both justified and feasible before significant resources are expended.

Proper governance will ensure that a program is justified through the development of a structured go/no-go decision process. The best technique for go/no-go decisions is a cost/benefit analysis. Cost/benefit analysis is a comparative assessment of all the benefits that are anticipated from the program and all the costs to implement, execute, operate, maintain, and make use of the outcome throughout its physical life cycle. It should also include any impact benefits/costs, such as environmental, societal, or economic, that the program will have on its community. These will be more subjective in nature, as the program's worth, or in contrast its cost, will be based on the wants and needs of the community, or even society as a whole, as opposed to value being inherent to the outcome itself. New York State's mandate to provide a "sound basic education" is an example of such a benefit. A consequential cost might be the negative environmental impact construction of needed facilities will have on the landscape. Another type of cost to take into consideration is opportunity cost. Opportunity cost is the cost measured in terms of the value of the next best alternative that was not chosen. Opportunity cost is a key concept in economics, and has been described as expressing "the basic relationship between scarcity and choice"[17]. An example from one of my recent programs was the decision to build a science classroom wing as opposed to a new auditorium. The opportunity cost to the community of the science wing was the forgone auditorium.

Once a program is determined to be justified, the team must fully investigate if they can pull it off, that is, "Is it feasible?" Determining program feasibility entails many things, including asking:

- Is it technically possible?
- Can it be funded?
- Are sufficient resources available?
- Is the team ready?
- Are stakeholders on board?

If, in broad terms, these are affirmed, then they should be more fully investigated in a detailed program feasibility study. An example of a well-executed feasibility study is the one completed for the $3.2 billion Chicago Region Environmental and Transportation Efficiency Program (CREATE) [18]. To determine program feasibility the team developed the Systematic, Project Expediting, and Environmental Decision-Making (SPEED) strategy. This was a structured approach for vetting answers to the critical program feasibility questions noted above. It was also a process to prioritize and select the component projects based on their alignment with the program goals and desired benefits. The SPEED strategy is more fully explored in the case study at the end of this chapter.

3.5 PROGRAM CHARTER

The program charter is a statement of the scope, objectives, and participants in a program. It provides a preliminary delineation of roles and responsibilities, outlines the objectives and expected benefits, identifies the main stakeholders, and establishes the authority of the program manager. It is a formal document, signed by each member of the program team, and serves as a future reference of what was agreed upon. It also marks the formal approval to proceed with the program. It is usually a short and concise document that references more detailed plans to be developed later in the program management process. The program charter includes:

- Mission and vision statements
- Organization structure
- Roles and responsibilities
- Project delivery method(s)
- Funding
- Budget constraints
- Schedule constraints
- Quality expectations
- Safety standards
- Communications procedures

The program charter is created through a series of brainstorming sessions. These sessions first establish the program community and working agreements. From there a high-level roadmap is developed that includes program scope and boundaries, the commitment of resources, and measures of success. The final chartering session includes a formal signing of the program charter detailing the items listed above. To be most effective, chartering sessions should include individuals from the program team with diverse skill sets and be conducted in an atmosphere that encourages creative, open-minded thought.

As a minimum, the chartering sessions should include the owner, the program management team, the design professional, the construction professional, the fiscal advisor, and the legal team. It's been my experience that project sponsors are sometimes reluctant to bring the whole team together this early in the process. This is a mistake, as the program charter will set the path for all future actions, and thus initial buy-in from the whole team is essential. There is a popular saying in construction, "Plan to build, and then build the plan." It's critical that the program participants who "planned to build" during the initiation process are the same ones "building the plan" during the execution phase. I liken this to the successful strategy that Kiewit employs with estimating. Instead of having a professional estimating department as do similar size construction firms, Kiewit uses its field managers to estimate and bid projects. If they are low bidder, the estimating team becomes the field team and becomes responsible for successfully executing the project. So during construction if there are issues with the estimate, there is no one to blame but themselves, and this encourages accountability.

In addition to the accountability benefit, having a diverse field of the right professionals will ensure effective interaction during these initial brainstorming sessions. Having the experience of a construction and legal professional, for example, in addition to that of the design professional in establishing the basic boundary of program duration, will result in more realistic and obtainable goals.

I know from experience that it can be a challenge to get good work out of such a diverse group of professionals. As Sigmund Freud observed, "Groups bring out the best and the worst in people." To bring out the best in the program team, the environment of the chartering sessions should be:

- *Free from fear.* Essentially this comes down to ensuring the chartering sessions are conducted in an atmosphere where participants are not afraid of saying something "dumb." Ideas should not be rated, ranked, or rewarded, nor should they be "corrected," chastised, or penalized.
- *Competitive.* This may seem contradictory to the above, but in the best chartering sessions, the program team should feel pressure to show off what they know and how skilled they are at building on others' ideas.
- *Structured.* In Chapter 6 of his book *Good to Great**, Jim Collins observed that "Creativity dies in an undisciplined environment." Proper governance of the charting session is essential if you are going to get the best out of everyone. I believe the best chartering sessions are facilitated by a construction program manager with exceptional emotional intelligence.

3.6 IDENTIFYING RISK

A critical part of the initiation process is to identify major risks and their likelihood of occurrence. Program risks are uncertain events or conditions that, if they occur, could have a positive or negative effect on at least one important objective. A risk may have one or more causes and, if it occurs, one or more effects.

---

* Collins, James. 2001. *Good to Great.* New York: Harper Collins.

Program feasibility cannot be fully determined without first determining the likelihood of major risk events. The most common program level risks are:

- Schedules that are too aggressive
- Inaccurate cost estimates
- Poor program management
- Poor project management
- Scope creep
- Lack of resources
- Unforeseen conditions

It is important, prior to the go/no go decision, that the program team discuss the likelihood of a major risk event, or combination of events, occurring. The team should feel comfortable that if a major risk were to occur corrective action could be taken to avoid program failure. By addressing the major program risks in advance of the planning process, the team can proceed with their eyes wide open.

3.7 CASE STUDY OF AN EFFECTIVE CM INITIATION PROCESS

An effective initiation process puts a program on the right track to realize its objectives and to achieve the desired benefits. As we explored in Chapter 1, the Big Dig is an example of the negative consequences that can occur if the right amount of focus and scrutiny is not carried out early on in the program management process. I would even take this a step further, and say that it is irresponsible for a program to be allowed to forge ahead without being properly vetted for both justification and feasibility, especially where public monies are involved. The Chicago Region Environmental and Transportation Efficiency Program is an example where the right focus and scrutiny during the initiation process has put a large and complex construction program on the right track to achieve its desired outcome and benefits.

Roughly one third of all rail freight in the United States originates, terminates, or passes through Chicago. Chicago is by far the busiest rail freight gateway in the United States. The city handles more than 37,500 rail freight cars each day and by 2020 that number is expected to increase to 67,000 cars per day [19]. CREATE will help both the major railroads and the Chicago area cope with this sharp increase in freight volume, while concurrently producing substantial improvements for motorists and rail passengers. CREATE is a public–private partnership among the State of Illinois, City of Chicago, private freight railroads, Amtrak, Metra Commuter Rail, and the US Department of Transportation.

Much of Chicago's railroad infrastructure is a century old and due to the increase in rail freight traffic, congestion has resulted in delays, highway congestion, air pollution, safety concerns, and interference with intercity and commuter trains. It is estimated that the cost of this congestion is $11 billion annually [20]. The Blizzard of 1999 [21] was the catalyst for CREATE as it exposed these fundamental flaws in the region's rail infrastructure system. Railroad trains in the storm's center were stalled or delayed 12 to 24 hours. Following the aftermath, the Association of American Railroads created the Chicago Planning Group (CPG) to study and provide solutions to the rail congestion issues for both passenger and freight services in the region. CPG identified several of the operational inefficiencies that contributed to the halt of service during the aggravated storm conditions. Soon thereafter, CPG established the Chicago Transportation Coordination Office (CTCO) to develop managerial solutions to the identified operational problems. CTCO developed a color-coded model to illustrate both passenger and freight traffic in the Chicago region and to show congestion levels under different scenarios. This model communicated to public agencies the public effects of rail service and the importance of establishing a more reliable system. In the years that followed, other organizations were created that led to the establishment of the CREATE program. An early critical component of CREATE was to develop a rail network simulation model that could test proposed improvement scenarios.

Upon announcement of CREATE the program team began meeting with elected officials at each level of government. Meetings were held with civic and business organizations interested in freight issues. The program team also reached out to stakeholder groups that would benefit from CREATE. This included a large audience as the program promised a "stronger regional and national economy" and a "better quality of life for Northeastern Illinois." Public presentations were conducted for any interested parties. These efforts culminated in the formation of the program charter, officially called the "Joint Statement of Understandings," which then led to the CREATE feasibility plan being issued in August 2003.

At the onset, the CREATE program identified over 100 projects as critically needed rail improvements. Because of funding constraints it was known early on that not all of these projects could be included in the program. The entire program initially was estimated to cost $1 billion, however, by the close of the initiation process that increased to the current budget amount of $3.2 billion. This was considered by the team to be the absolute minimum amount to move forward while maintaining the integrity of the program's mission and vision. The railroad partners agreed to provide $230 million with the remainder to come from federal, state, and local government sources. Even at $3.2 billion, enough funds were not available to do all the projects, so the program team developed an innovative screening process called the Systematic, Project Expediting, and Environmental Decision-Making strategy. The SPEED strategy was designed to prioritize and expedite projects while continually assessing program justification and feasibility. I have summarized the strategy in the Figure 3.3.

The SPEED strategy addressed the CREATE program by supporting systematic decision making as well as providing an expeditious method of moving low-risk component projects forward. As described by the program manager, William C. Thompson of the Association of American Railroads, the answer lay in "A knot of interrelated problems requires

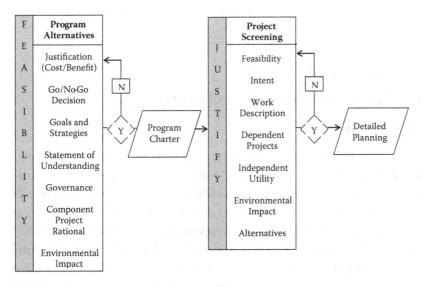

FIGURE 3.3
SPEED strategy.

a network of solutions." So under the SPEED strategy, alternatives are vetted first to determine justification for, and then the feasibility of, the program itself. For the component projects it is just the opposite: feasibility is established first and then justification is determined. This is the case because once the path for the program is chosen, the decision process becomes "how to get there." Projects are planned and selected based on specific technical requirements, such as "improving train speeds from an average of 9 miles per hour to 15 miles per hour." These projects fell into three categories: rail–highway grade crossings, viaduct improvements, and railroad infrastructure projects. Projects are then further prioritized by a variety of methods. For viaduct improvements, a survey of viaduct conditions is used to identify and prioritize the projects. A previous study identified the most congested rail–highway grade crossings. Computerized modeling was utilized to measure the severity of existing rail chokepoints in the region to prioritize rail–highway grade crossings and railroad infrastructure projects.

With more than a dozen of the projects in place as of early 2012, rail officials say they have already seen some reduction in delays with bigger improvements to come, according to Joe Shachter, Director of Public and Intermodal Transportation for the Illinois Department of Transportation. "The next two or three years in particular we think are going to show great advances," he said [22]. Given the great deal of energy and focus put into the initiation process, I have no doubt this will be the case.

3.8 CHAPTER SUMMARY AND KEY IDEAS

3.8.1 Chapter Summary

In this chapter the importance of having the right level of energy, focus, and direction during the critical CM initiation process was demonstrated as well as the need not only to justify the program, but also to determine that it is feasible. This must be done before significant resources are expended. Not doing so would be irresponsible with possible irreversible negative consequences. Determining program feasibility involves defining the ultimate purpose, skillful governance, and acknowledgment of uncertainties that can derail the endeavor. Ultimately determining program feasibility results in a go/no-go decision.

3.8.2 Key Ideas

1. The initiation process is where the collective ideas of the program team are channeled and a path is chosen to achieve success. The program initiation process requires both strategic and tactical thinking and the unique skills of a leader. The initiation process will result in the decision to proceed, or not, to the planning phase. The output of a successful initiation process is the program charter which outlines the high-level roadmap for the program to realize its benefits.

2. The ultimate purpose of the program must be agreed upon, and committed to, by the team. This is accomplished by defining the mission, articulating the vision, and developing a strategy.

3. A program must be justified and proven feasible before it can be permitted to proceed to the planning process. Proper governance will ensure that a program is justified and feasible through a structured go/no-go process. Things considered include: is the program technically possible, can it be funded, are sufficient resources available, is the team ready, and are the stakeholders on board. Identifying and assessing the likelihood for major risk events is also a key component.

4

Planning Process

4.1 INTRODUCTION

My wife, often to my consternation, uses the adage of the 5 Ps of success "Proper Planning Prevents Poor Performance." This catchy modern phrase probably owes its origin to Benjamin Franklin who, in his *Poor Richard's Almanac*, wrote "By failing to prepare, you are preparing to fail." Both sayings are ways of expressing the importance of being proactive and not passive about preparing and organizing for proper planning for the future. For construction programs proper planning is an absolute requirement for success.

4.2 PROGRAM MANAGEMENT PLAN

During the initiation process program options were vetted, a shortened list of scenarios was investigated for justification and feasibility, and the best path forward was taken. The program mandate and architecture were established, and resources were committed. This culminated in the development of the program charter that was formally adopted by the stakeholders which imparted authority to the program manager to proceed on to the planning phase. The next step in the process is the development of the program management plan. The program management plan is developed to accomplish the program outcome and ultimately to realize the targeted benefits, guided by the high-level roadmap that was developed during the initiation process. The program management plan is a stand-alone document, separate and different from the project management plans that are required to manage the individual projects within the program. In contrast to the planning for the program's projects, the program management plan typically is not developed through a series of iterations. Instead, the planning effort involves the evaluation of the project plans for conflicts or misalignment. The ultimate goal is to ensure this alignment, and produce a concise consolidated view of all program work, budgets, timeframes, and desired project outputs and program outcomes. The program management plan will set the monitoring and control structure for the strategic goals and the project plans will set the monitoring and control methods for the tactical goals. As such, the program management plan is not utilized to direct work and allocate resources. That is the purpose of the project plans. As these constituent plans are further developed, analyses and evaluations are conducted to ensure they are in alignment with the programwide goals. This enables managers to assess the program's progress against the plan and detect potential problems and take corrective action if needed.

Components of the program management plan include:

- Creating the master schedule
- Creating the master budget
- Developing the safety plan
- Developing the quality management plan
- Developing the risk management plan
- Developing the change management plan
- Developing the communications management plan
- Establishing rules of engagement
- Developing the transition plan

Components are developed in a collaborative way with input from each member of the program team. The program manager is responsible for coordinating efforts resulting in a synchronized formal document containing comprehensive plans for each element. When complete, the program management plan will be used as the basis to manage all construction activities during the execution process.

4.2.1 Creating the Master Schedule

In construction, the most critical planning task is the creation of an effective work schedule. Because of the potential for squandering scarce resources, it is irresponsible to begin any level of construction activity without a well-defined and organized schedule. This applies to the most basic of things such as a foreman preparing for a day's construction work to complex tasks such as a program manager planning for the execution of multiple interrelated projects spanning several years. Because of the complex nature of programs, the work plan must be initiated with the establishment of a detailed master schedule. Not widely understood is that other program boundaries, such as budget, level of quality, and scope, cannot be fully determined without first establishing the program timeframe. A good example of this is the 3,200 square feet (sf) home (see Figure 4.1) and 2,800 sf boxing facility we built in 170 hours for ABC's *Extreme Makeover: Home Edition* in Geneva, New York [23].

Although the cost must remain undisclosed, it was an "order of magnitude" greater than what a similar project would cost had it been done at a regular pace. This may seem like common sense, but I have seen many program managers establish a budget without first determining the schedule. And the setting of the program timeframe can only be correctly done through the development of a detailed master schedule.

The state-of-the-art technique for creation of detailed construction schedules is the critical path method (CPM). The CPM approach was developed in the 1950s by the US Navy [24] and is a mathematical algorithm for

FIGURE 4.1
Extreme makeover home in Geneva, New York.

scheduling a set of interrelated activities. The technique has been refined in recent years to include resource-leveling, which prioritizes activity sequencing based on the scarcity of resources. Under the CPM approach, schedules are developed by first creating a model of the program using the following:

- A list of activities required to complete the program
- The time required to complete each activity based on the availability of resources
- Established milestones
- The dependencies between the activities

Based on this information the CPM algorithm calculates:

- The overall program duration.
- The earliest and latest that each activity can start and finish without delaying the program. This is referred to as float time.
- The activities on the "critical path," which if delayed will cause a direct delay to the program completion date. Activities on the critical path are said to have no float time or "zero float."

CPM software products such as Primavera P6® and Microsoft Project® can create these models for programs with thousands of activities, determine the crucial path for the desired outcomes within minutes, and are commercially available for under $2,000.

The most challenging part of the CPM scheduling process is the development of the network diagram which is a graphical representation of the schedule. Creation of the network diagram requires a collaborative effort from the program team, best done through brainstorming and what-if scenario planning. Developing the network diagram or what is sometimes referred to as the *schedule model*, is a highly complex and demanding task. Essentially the team will need to work backward from the completion date and develop a plan of activities that results in the expected outcome for the program. As Sherlock Holmes noted:

> Most people, if you describe a train of events to them, will tell you what the result would be. They can put those events together in their minds, and argue from them that something will come to pass. There are few people, however, who, if you told them a result, would be able to evolve from their own inner consciousness what the steps were which led up to that result. This power is what I mean when I talk of reasoning backward [25].

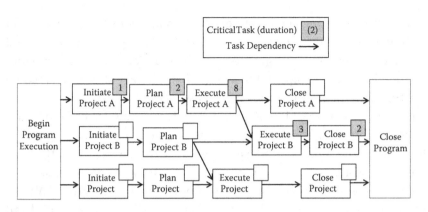

FIGURE 4.2
Network diagram.

Because programs must realize their desired outcomes (and benefits) within a defined timeframe, the network diagram must be developed in a similar way, through backward reasoning. A simplified example of a network diagram is illustrated in Figure 4.2.

All network diagrams for construction programs include five critical elements: the activity list, milestones, task durations, dependencies (predecessors and successors) between tasks within a project, and the dependencies between tasks from other projects within the program. The latter is critical, as can be seen from the simplified example in the figure, where Project C has the longest overall duration and all three projects start at the same time. But because of dependencies between projects, none of Project C's tasks are on the program's critical path.

The most critical element in developing the network diagram is the activity list. Program activities selected to be included in the detailed master schedule must meet the following criteria:

- Tasks must be specific to a program management process group.
- Performance of the task must be critical for the proper execution of the program.
- Tasks must be temporary with a determinable duration.
- Tasks must be able to be defined in a meaningful way and understood.
- A task's progress must be measurable.

The best technique for developing the activity list is to start with a *program work breakdown structure (PWBS)*. The PWBS is a treelike structure delineating the hierarchical components of the program and setting them into the program management process groups, projects, and ultimately program packages from which individual tasks or activities to complete them comprise the activity list. The PWBS is developed by starting with the end objective and successively subdividing it into smaller and smaller manageable components in terms of size, duration, and responsibility. A simplified and abridged example is shown in Figure 4.3 to illustrate the concept. The shaded boxes are the lowest level of the hierarchy and would be included in the detailed master schedule as tasks.

One of the most challenging parts of developing an effective PWBS is determining how far to drill down so that the appropriate level of detail is provided for each task on the activity list. It is also important that not too much detail is provided which can make the schedule unmanageable. I used to tell my team that "You get no points for making it more difficult." Determining the right level of detail is more of an art than a science. It is subjective, requiring consideration of the following:

- The importance or priority of the work
- The ability to accurately determine a fixed duration

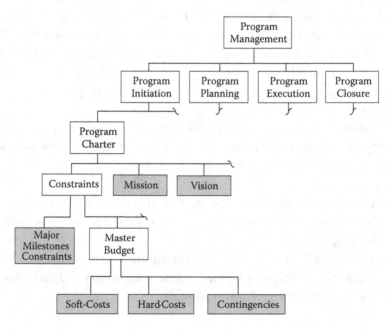

FIGURE 4.3
Program work breakdown structure.

- The proficiency of assigning dependencies with other activities
- How the activities will be used for monitoring and control purposes

Using these criteria, activity durations are set based either on the availability of resources (effort driven) or from anticipated, or required, production rates (task driven). Activities are generally resource driven during the initiation, planning, and closure processes. This is the case because select individuals or teams must perform these tasks, and thus the timeframe is based on the availability of those involved and level of effort required by the specific individual or team. For example, for a complex bridge renovation program, the availability, and effort required, of a particular structural engineer, who is proficient at designing structural steel components, may determine the duration of those tasks.

Activities are generally task driven during the execution phase. For construction, production rates based on standardized crew compositions have been well established and documented for almost any type of work. Publications such as the *RSMeans® Cost Data Books* contain these data which can be adjusted for factors such as site location and type of construction. Many construction companies also retain their own production data from past programs for this purpose, and for estimating costs.

In contrast to resource-driven activities, task-driven activity durations can be manipulated to meet schedule requirements by adjusting the level of resources. For example, if the program schedule requires that the construction of a project component be accelerated, adjusting the crew size would be an option. This is not the case with resource-driven activities. For example, having the program manager do a week of "double shifts" to catch up on planning activities would not be a good strategy. The consequence of this distinction between resource- and task-driven activities is critical to remember when creating the schedule.

Milestones establish the schedule boundaries and constraints for performing the work on the program. The major program milestones, including the start and finish dates for each critical phase, are established during the initiation process, and intermediate milestones are defined and set during the planning process.

For all programs there is an ideal time to start and finish each phase of the work. The major milestones are inputs to the scheduling process. Under the best of circumstances, it will be within the team's power or influence to set the major milestones based on what is best for the program and its stakeholders. Under that scenario, items such as the readiness of the team, the availability of resources, or economic conditions, may play a role in setting the major milestones. On some programs, such as the LIPO program, the team does not have this flexibility, as the major milestones are mandated.

Intermediate milestones are an output of the scheduling process. Intermediate milestones are primarily used as an accountability tool, to monitor and control progress and adherence to the plan. Examples of intermediate milestones from some recent program master schedules are:

- Buildings weather-tight prior to winter.
- Begin start-up and testing of effluent pumps.
- Complete Bridge 6B.
- Begin conceptual design.
- Acquire permits.

Once the list of program tasks is established, durations are set, and milestones are confirmed, the next step in the CPM scheduling process is to assign dependencies between tasks. Dependencies are often referred to as "logic ties" and are assigned based on the relationship between predecessor or successor tasks. A predecessor task is one whose start or finish date determines the start or finish date of a following task. A successor task is one whose start or finish date is driven by its predecessor task.

Assigning dependencies takes considerable knowledge, experience, and judgment. It requires an understanding of how construction activities are sequenced and how resources are shared across the program. It is an iterative process where different sequencing scenarios are examined until the most efficient solution is found. This is best illustrated by example. Figure 4.4 shows a fragnet, or a portion of a master schedule, from the $34 million, Bridge 25, Interstate 84/County Route 8 Interchange program in Waterbury, Connecticut. Hammock, or summary level tasks have been shown for clarity.

The renovation work on the bridges included both hydrodemolition[26] with repair work and the complete reconstruction of the structure. The bridges were required to be completely closed to traffic or have travel lane restrictions, depending on the nature of the renovation work. For roadway programs, travel lane restrictions require what is referred to in the trade as "maintenance of traffic" measures, so the acronym of MOT is used in the figure where it applies.

The sequence logic for the project of renovating Bridge 25A is shown in its entirety in Figure 4.5. Each task is a physical prerequisite to the next

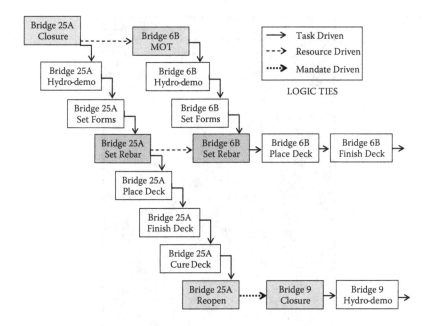

FIGURE 4.4
Activity sequence logic.

so creating the sequence logic diagram for this particular project was straightforward. In contrast, Bridge 25A's logic ties with other project tasks within the program were not driven by physical constraints, rather they were driven by resource requirements. For example, because of the limited availability of skilled ironworkers in the Waterbury area, only one of the bridge decks could have reinforcing steel (rebar) installed at any one time. Similarly, a specialty composite crew was responsible for MOT operations, including both the closure of Bridge 25A and setting up and dismantling of the traffic pattern for Bridge 6B. For Bridge 9, the dependency with Bridge 25A was not task driven or driven by resource requirements, but rather by a program mandate. It was agreed, in an effort to minimize traffic disruption, that no more than one bridge could be closed at a time. Because both bridges required closure for their respective scopes of work, the completion of Bridge 25A turned out to be a prerequisite to the start of work on Bridge 9.

The fragnet in Figure 4.4 shows the complexity involved in properly sequencing just a small portion of a complex construction schedule. The completed master CPM schedule totaled over 2,500 activities with over 3,500 task dependencies. Sometimes the challenge of creating

FIGURE 4.5
Roadway interchange in Waterbury, Connecticut.

the CPM schedule can seem overwhelming at first. I know that was the case for me on the Waterbury schedule. But if a systematic approach is taken, starting first with the PWBS, and then tackling the program, "one project at a time," a good approach and plan can be developed. For the Waterbury schedule we used a rather "nontech" approach of sticking over 2,500 Post-it® notes with task descriptions on the job trailer's walls. As we discussed "what-if" scenarios, we would physically move the tasks to align with the revised sequence logic. For each scenario we would then revise the network diagram on scheduling software and then run it to calculate the critical path and completion date. We repeated that process until we had the best plan of attack for reconstruction of the 25 bridges.

4.2.2 Creating the Master Budget

The team committed to the basic budget guidelines as part of the process for developing the program charter. The next step in the planning process was a detailed master budget to further allocate the budget into discrete cost categories. The main purpose of developing the detailed master budget is to create a comprehensive inventory of all program costs, which then can be used as a baseline for monitoring and controlling expenses. Depending on the program, it also might be critical to allocate the budget according to funding source(s). This is especially true for public programs, where different funding sources may require different criteria for reimbursement. Once established, the comprehensive inventory of cost is then divided into the three major budget categories. The major budget categories are indirect costs, direct costs, and contingencies.

Indirect costs are costs that are not directly accountable to a program component (i.e., planning) or a specific constituted project (i.e., a specific public outreach program). For construction programs a significant portion of indirect costs is associated with the *program management office* (*PgMO*). The PgMO is the governing body responsible for developing and managing the constituted projects and program components. In addition to the cost of establishing and maintaining the PgMO's staff, other indirect expenses such as long- and short-term borrowing costs, insurance premiums, legal fees, and the like, might be included in the *indirect cost* category, depending on the nature of the program.

Conversely, a *direct cost* is an expense that can be directly attributed to a specific program component or constituted project. In construction, direct expenses are further distinguished as either soft or hard costs. *Soft costs* include items that will not become a physical (i.e., "hard") part of the completed project. Soft costs are also referred to in the trade as "incidental expenses" or "extraneous costs." Numerous standardized lists of soft cost categories, and templates, have been developed for various types of construction programs. Table 4.1 is an example of a standardized list for K–12 school projects in New York State. I like this particular template as it also includes the responsible party for determining each cost. Regardless of the origination of the template, it is imperative that the team develop a comprehensive list of soft costs that is specific to the program. This will ensure that soft cost items (or categories) are not "missed" and become unbudgeted expenses.

Hard costs are expenses for items that will become a physical part of the completed program. In common practice, there are generally two methods to determine hard costs in construction: analogous and parametric

TABLE 4.1

Soft Cost Template

Resp	Budget Item	Amount	Totals
A	**Architect Fees**		
	Architect Base Fee		
	Architect Reimbursables		
	Architect Additional Services		
C	**CM Fees**		
	Clerk of the Works		
	Construction Management		
S	**Legal Fees**		
	Legal Costs (Contracts, Disputes, etc.)		
	Bond Counsel		
	Financial Services		
	Financial Consultant Fees		
	Bond Costs		
A	**Site Investigation and Survey**		
	Land Survey		
	Soil Borings		
	Topographic Survey		
	Asbestos Survey		
	Stormwater Management Plan		
C	**Special Testing**		
	Concrete Testing		
	Geotechnical Testing		
	Asbestos Monitoring		
	Steel Testing/Special Inspections		
	Lead Paint Testing		
	Other (Spray F.P., Asphalt, etc.)		
S	**General Administration**		
	Printing During Design		
	Printing for Bid Sets		
	Printing for Construction		
	Postage		
	Telephone Costs		
	Public Relations		
	Bond Costs		
	Moving Costs		
	Cleaning Costs		
	Construction Field Office Costs		
	Temporary Utilities		

TABLE 4.1 (*Continued*)

Soft Cost Template

Resp	Budget Item	Amount	Totals
S	**Insurance Premiums**		
	Builder's Risk		
	Site Purchase		
	Site Acquisition Costs		
	Site Development		
	Site Development Costs		
	Utilities and Services		
	Sewage Work		
	Water Services		
	Gas Service		
	Electric Service		
	Telephone Service		
	Fire Alarm		
	Furniture, Fixtures, and Equipment (FF&E)		
	Computers		
	Furniture and Other		
	Design and Specify FF&E		
	TOTAL AUTHORIZED BUDGET		

Note: Responsible codes: A: Architect, C: Construction Manager, S: Owner.

estimating. *Analogous estimating*, often applied by architects and other design professionals, uses similar past projects, and thus the root of the word: *analogy*. In its simplest form analogous estimating can be used to estimate a project's hard cost based on general information such as the building type, location, and time of construction. The information can be gleaned from the design professional's internal cost records of past projects, or it can be obtained from comprehensive cost databases provided by online services such as Marshall and Smith®, cost reference books such those published by Reed Construction Data® in the RSMeans series, or the Craftsman® series of cost manuals. An example of analogous estimates for different building types in Syracuse, New York, for 2010 [27] is shown in Table 4.2.

In contrast, *parametric estimating*, often used by construction managers and contractors, uses the relationship between parameters (or variables) to calculate cost, thus the root of the word: *parameter*. Parametric estimating is applied to develop a thorough task list, and then detailed cost estimates,

TABLE 4.2

Analogous Cost Estimate

Building Type	SF Cost ($)
Storage Warehouse	57
Industrial Building	63
Discount Store	75
Residence	89
Community Shopping Center	96
Small Apartment	97
Retail Store	113
Convenience Store	124
Motel	135
Office Building	135
Day Care Center	148
Medical Office	177
Fast Food Restaurant	201
Bank	309

TABLE 4.3

Parametric Cost Estimate

ITEM	UNIT	QTY
Steel Columns	Ea	48
Steel Beams	Ea	200
Total Weight	Tons	147
Average Weight	Lbs	1,200
Production	Tons/Day	30
Duration	Days	5
Crew Costs	$/Day	$33,000
Total Erection Costs	**$165,000**	

of each scope item (see Table 4.3). When there is limited information, this is achieved by envisioning "building the project on paper," as illustrated in Figure 4.6 for steel erection, and then applying parameters, such as equipment, material, and labor unit costs, to the quantity of each discrete work item. This requires a program team with a great deal of experience and construction know-how but if applied correctly can lead to accurate results.

For example, we used parametric cost estimating on the Cayuga-Onondaga BOCES program to develop a detailed cost estimate during

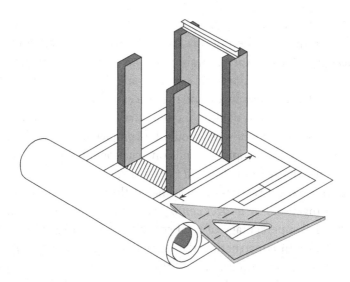

FIGURE 4.6
Building the project on paper.

FIGURE 4.7
Cayuga-Onondaga BOCES.

the conceptual design phase. Through parametric estimating, we were able to drill down to the "anchor bolt level" even though the architect had not even started the structural drawings yet. That conceptual estimate turned out to be within $3,200 of the actual program cost of $42.5 million (see Figure 4.7).

There is debate among the professions regarding which method of cost estimating produces the best results. In my opinion, the best approach is to have a design professional perform an analogous estimate and a construction professional perform a parametric estimate, and then reconcile the results. Each method has its advantages, and the process of reconciliation itself can lead to a healthy exchange of ideas and potentially more accurate estimates. For the best results, the process should be completed at each design gate (conceptual, design development, and construction documents) and as the design becomes more refined so will the estimates.

Once indirect and direct costs are estimated, the team will need to develop budget contingencies. A *budget contingency* is a predetermined amount, usually fixed as a percentage of total cost by budget category, to account for uncertainty. In construction there are three general types of budget contingencies: program, design, and project. The program contingency is a reserve set aside for risk responses. Setting of this contingency will be based on several factors, including the probability of the occurrence of the risk, the potential for the manifestation of multiple risks, and the potential for concurrent risk events. A design contingency allows for the fact that projects often contain more elements when they are fully designed than could have been anticipated earlier in the design process. As such, the contingency will reduce as the design becomes more defined as it passes through its milestone gates (conceptual–design development–construction documents). The project contingency is for unknowns during construction. The project contingency allows for unknown factors, such as unforeseen site conditions (i.e., discovery of an underground petroleum storage tank), that could increase construction and related costs beyond the estimate. Establishing these contingencies is subjective in nature and will depend to a great extent of the type of program, the project delivery method(s), and ultimately the owner's risk tolerance.

4.2.3 Developing the Quality Management Plan

The *quality management plan* describes the program's strategic approach to ensure the delivery of high-quality construction projects and ultimately the program's benefit(s) to the stakeholders. The *quality assurance process* is concerned with establishing the authority of the quality assurance function and processes for monitoring and evaluation of quality in relation to established standards. Quality assurance activities concentrate on the prevention of problems through the continuous improvement of processes.

The first step in quality planning is establishing a programwide quality assurance organizational structure. All major construction programs should have a dedicated quality assurance manager who has the sole responsibility, and the critical authority, to govern and implement the quality assurance functions. A basic imperative for implementing quality assurance programs is that the quality assurance manager must report directly to upper management. To perform his or her quality assurance evaluations and implement quality control functions, the quality assurance manager must also have direct access to the program manager and each project manager. A composite organizational structure allows this direct access to the highest and lowest organizational levels at the same time.

The next step is to establish programwide quality standards. Standards for construction quality are issued in publications of organizations such as the American Society for Testing and Materials (ASTM), the American National Standards Institute (ANSI), or the Construction Specifications Institute (CSI). Technical specifications have also been developed for particular types of construction work, such as welding standards issued by the American Welding Society, or for particular project types, such as the Standard Specifications for Highway Bridges issued by the American Association of State Highway and Transportation Officials. During design, these general specifications are adapted to reflect federal, state, and local building code requirements, local conditions, available materials, and other circumstances specific to the program or component project.

Quality management planning must consider cost/benefit tradeoffs. The primary benefit of meeting quality standards is less rework and the primary cost is the expense associated with quality management activities. In construction, even minor defects may require rework and in the worst case, failures may cause personal injuries or fatalities. Because the cost of failure is so great, set procedures will be mandated in the technical specifications for construction materials testing and the inspection of workmanship. Construction is unique in that a majority of the program work is performed by outsourced contractors and vendors. Quality assurance through materials testing and workmanship inspection are essential, as they provide the evidence to establish confidence that the materials and techniques used by construction contractors are in compliance with the general intent of plans and specifications. The quality management plan will determine what materials will be tested, the frequency of the tests or inspections, the physical location for testing, and the entity responsible and accountable for the results. For most construction programs, materials testing will include, but not be limited to, the following:

- Soil stabilization inspection and compaction
- Concrete mix design and batch plant inspection
- Cast-in-place and precast concrete placement inspection
- Structural steel inspection (welding, bolt torque, and fabrication)
- Asphaltic mix design, batch plant inspection, and pavement evaluation
- Roofing, fireproofing, and paint inspection
- Masonry inspection and testing

Unlike construction materials testing, workmanship inspection is more of an art than a science. The goal is to make sure the contractor "gets it right the first time" and thus avoid rework. Rework is not only costly to the contractor but can also have a severe impact on the overall program schedule. Because of the concept of time is money, a relatively small project deficiency that requires rework can result in severe budget consequences for the entire program. The technical specifications developed by the design professional will contain the quality standards for the work and will become a component of the quality management plan. The technical specifications will be used as a guide for the design and construction professionals to inspect the contractor's work.

Determining if a contractor is capable of, and willing to, perform quality work is an important aspect of the quality assurance process. This is subjective in nature, and will take the skill of a knowledgeable inspector with experience with the specific type of work at hand. It is also important that there is full-time inspection of the work. Sometimes, in an effort to save cost, an owner will request part-time inspection, sometimes referred to in the trade as *construction observation*. This should be avoided at all cost. Full-time inspection is critical in avoiding rework, as it cannot be determined in advance when deficiencies will occur. The quality management plan should address workmanship issues by mandating that each project site is staffed with capable, full-time, construction inspector(s). The quality management plan should also provide guidance to the inspectors for grading a contractor's performance. Things to consider as guidance for grading a contractor's performance are:

- Overall management of work
- Construction method and techniques

- Ability to do the job
- Labor relations
- Worker attitudes
- Supervision: quality
- Supervision: quantity
- Home office support
- Control of subcontractors
- Cooperation with other contractors
- Cooperation with the program management team
- Attitude toward correcting errors
- Responsiveness
- Equipment availability
- Tool supply
- Job cleanliness and orderliness
- Cost control
- Scheduling control
- Quality control
- Safety program
- Environmental stewardship

It is also critical that the quality management plan puts in place procedures for assisting in selecting the right contractors to perform the work in the first place, and the right entity to monitor quality compliance. For most construction programs, both construction material testing and workmanship inspection functions will be outsourced. In fact, many building codes and technical specifications require outsourcing as a method to assure accountability and reduce the potential for conflict of interest. There are many excellent firms providing these services. Items to consider when choosing a firm are:

- *Stay Local.* Local construction or planning codes vary quite a lot from region to region, state to state, and country to country, so calling on the services of a locally based testing and inspection firm is often the best thing to do. Local firms will also be more agile and be better able to provide services on demand, if required.
- *Use a Quality-Based Selection Process.* For public programs in the United States, federal statute permits a quality-based selection process for these types of services. This is the best approach, as the most qualified firm should be selected regardless of the cost for services.

- *Screen the Inspectors.* As discussed, it is critical that the actual on-site individuals assigned to perform the quality assurance and control functions for each project are first thoroughly screened by the selection committee. To be effective, they must be competent, experienced, energetic, and most important, have a set of core values that are in alignment with those established in the program charter. Each individual should be required to review, and sign off on, the quality management plan, prior to assignment to a project position.

As far as selecting the right contractors, that will depend to a great deal on the project delivery method. For public work, where there is a requirement to "select" the low bidder, the quality assurance function is an even more crucial part of program success. Where this is not a requirement, establishing procedures for prequalification of all vendors and construction contractors can go a long way in ensuring success. In either scenario, providing high-quality bidding documents, which will establish the appropriate rules of engagement, will also play a critical role. More on that later.

4.2.4 Developing the Risk Management Plan

A critical part of the planning process is to identify, and then analyze, program risks as either threats or opportunities. Identifying and analyzing risks during the planning process will position the team to maximize the possibility and results of positive events and minimize the probability and consequences of adverse events. The process involves four elements:

- *Risk Identification.* Determining which risks may affect the program and defining and documenting their characteristics
- *Qualitative Risk Analysis.* Prioritizing risks for potential further analysis or action by assessing and combining their probability of occurrence and impact
- *Quantitative Risk Analysis.* Numerically analyzing the effect of identified risks on overall program objectives
- *Risk Response Strategies.* Developing options and actions to enhance opportunities and to reduce threats to program objectives

The first step in the process is to identify items that may have the potential to affect the program. Risk elements associated with schedule, scope, cost, and resources, in that order [28], are a good place to start for most programs. When identifying risks, the team considers:

- *Threats:* A risk that will have a negative impact on a program's objectives if it occurs (what might happen to jeopardize the program's ability to deliver its benefits). An example of such a threat would be the sharp increase in the cost of crude oil. For a major roadway program, this risk will have a serious impact on costs, potentially resulting in a reduction of asphalt paving scope.
- *Opportunities:* A risk that will have a positive impact on a program objective if it occurs (what might happen to improve the program's ability to achieve its benefits). If the price of crude oil goes down sharply, this could result in additional asphalt paving scope.
- *Residual:* Risks that remain even after developing responses to the original risks. An example would be the final cleaning crew's work on slippery floor surfaces. While the floor is still wet, a contractor employee could enter the area and slip and fall. To reduce the likelihood that this will happen, "caution" signs are placed on all wet floors after cleaning. Although the signs may reduce the potential for slips and falls, these accidents may still occur. This remaining risk is termed residual risk.
- *Secondary:* These risks are caused by responses to the program's original risks. For example, a general contractor is assigned the responsibility for site safety to mitigate the owner's exposure to workplace compensation claims. A secondary risk occurs as a result of using an external vendor as the general contractor, who, without proper oversight, could be negligent in their duties.
- *Interaction:* This is the combined effect of two or more risks occurring simultaneously that can be greater than the sum of the individual effects of each free-standing risk. For example, in the United States, federal budget cuts may increase delays in Federal Highway Administration permits, at the same time that federal programming dollars become scarcer. The combined effect of the permit delays (time is money), and less program funding, could seriously reduce the scope of a roadway program, potentially even making it infeasible.
- *Triggers:* These are symptoms or warning signs that indicate whether a risk is becoming a near-certain event and that a contingency or response plan should be implemented. For example, a trigger for a sharp increase in the cost of crude oil, and thus asphalt prices, might be escalation of unrest in the Gulf oil states.

After risks have been identified they must be analyzed to assess their potential impact on the program. There are two types of risk analysis: qualitative and quantitative:

- *Qualitative Analysis:* A qualitative analysis categorizes the identified risk sources and factors and then generally classifies them relative to probability of occurrence and severity of impact. For construction programs, the main categories of risk, in no particular order, are: economic (i.e., a sharp increase in crude oil prices), technical or operational (i.e., mechanical process issues with a state-of-the-art wastewater system), organizational (i.e., poor program team compatibility), political/societal (i.e., the 9/11 terrorist attacks), environmental (i.e., a SEQR challenge [29]), and project related (i.e., discovery of an unforeseen site condition). Categorizing aids in analyzing risks as well as planning responses or contingencies for them.

In qualitative analysis, the classification of risk is largely subjective in nature, and usually is done through interviews or brainstorming with the program team. This analysis relies heavily on the team's experience, wisdom, and judgment. For example, it should be clear from recent experience, that there is potential for a sharp increase in crude oil prices and this could have a severe impact on a roadway program. Thus, for this type of program, this would be a priority risk to focus on during the execution phase. A detailed response plan should be established, such as including an asphalt material escalation clause in the construction contracts.

Once the information is gathered regarding risk probability and impact, simple graphic techniques, such as shown in Figure 4.8, can help summarize and illustrate the priority of each risk.

- *Quantitative Analysis:* During qualitative analysis, it may become appropriate to enter into a more detailed quantitative analysis. This may result because of the potential severity of the impact or the difficulty in determining the likelihood of the event. Quantitative analysis will enable the impacts of the risks to be quantified against the basic project success criteria: cost, schedule, and performance. Several techniques have been developed for analyzing the effect of risks on the final outcome of programs. In practice, these techniques

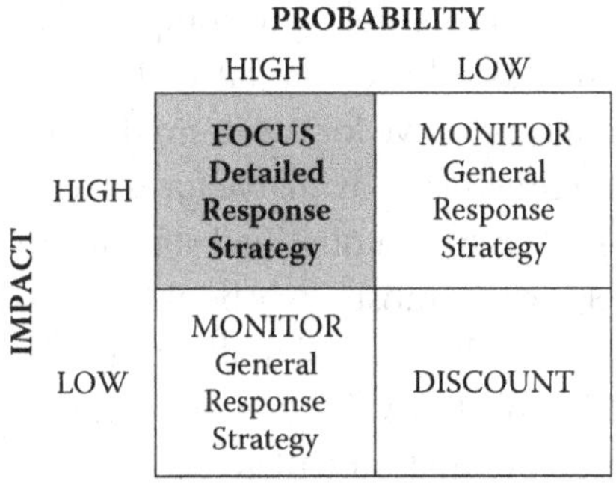

PROBABILITY

		HIGH	LOW
IMPACT	HIGH	**FOCUS** **Detailed** **Response** **Strategy**	MONITOR General Response Strategy
	LOW	MONITOR General Response Strategy	DISCOUNT

FIGURE 4.8
Quantitative risk analysis.

focus on the likelihood of attaining the schedule and cost goals and typically include sensitivity analysis, probabilistic analysis, influence diagrams, and decision trees.

Sensitivity analysis is often considered to be the simplest form of quantitative risk analysis. Essentially, it determines the effect of variations in just one risk factor, such as incremental increases in the cost of materials, on the outcome of the program. Typically, only adverse changes (threats) are considered in sensitivity analysis. Sensitivity analysis is carried out as follows: a priority risk is identified, a program performance measurement is set (usually completion date or budget), a range of possible values for the risk factor is set (i.e., crude oil prices of $75/barrel through $150/barrel), the impacts on program performance are determined, and the results are either plotted or tabularized. The results will help determine which risks have the greatest influence on program success, and how sensitive program outcomes are to variations in the risk factor. It may also determine the "tipping point" where the program becomes infeasible. *Probabilistic analysis* specifies a probability distribution for each risk and then considers the effect of risks in combination. This is perhaps the most common method of performing a quantitative risk analysis and is the one most construction professionals consider, incorrectly, to be synonymous with the entire risk identification and analysis process. For construction programs, the most common form of probabilistic analysis is referred to as Monte Carlo simulation. Monte Carlo simulation is carried out as follows. Priority risks are identified, a program performance measurement is set, the uncertainty for the risk is established (each is given a best, most likely, and worst case value), and a model is run using computational algorithms that rely on repeated random sampling to compute the result. Monte Carlo software products, such as @Risk, Crystal Ball, Risk+, and Decision Pro [30], can create a model and solve for the desired outcome within a matter of minutes and are commercially available for under $1,000. Whereas the result of a sensitivity analysis is a qualified statement ("If crude oil prices stay below $95/barrel, we will most likely be able to do the full scope of the paving program"), the result of a probabilistic simulation such as Monte Carlo is a quantified probability ("If we build the dam, there is a 20% chance that the salmon population will go extinct").

Influence diagrams and *decision trees* are graphical methods of structuring models of possible outcomes. They show the present possible courses of an action and all future possible outcomes. In construction; these forms of risk analysis are often used to determine the probability of meeting the program budget. Analytically, they provide a common framework for a side-by-side comparison of alternative strategies. Once drawn, they can be used to communicate uncertainties that affect the ability of the program to create value or deliver its benefits fully. Figure 4.9 is an example of a simple influence diagram for determining the feasibility of meeting the budget on a roadway program. An arrow denotes an influence, an oval denotes a risk factor (an uncertainty), and a rectangle

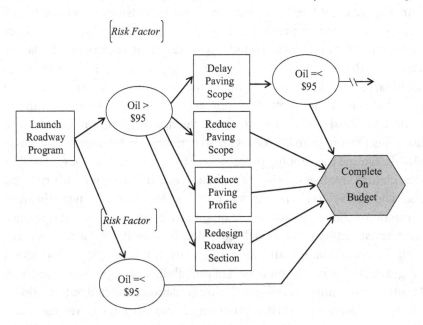

FIGURE 4.9
Simplified influence diagram. (Please see Note 31.)

denotes a decision. In practice, these diagrams can include hundreds of risk factors and possible outcomes. The value in just the process of creating these diagrams for team collaboration, program understanding, and buy-in, cannot be understated. It's like the old adage, "The journey is the destination."

In order to determine program feasibility fully, it is critical not just to identity and define risks but also to know that practical responses can be developed for all potential scenarios. Once risks are identified, defined, and analyzed, options and actions must be developed for the various potential outcomes. During this process, critical things to consider include:

- Risks must be addressed by priority.
- Responses must be appropriate to the severity of the risk.
- Responses must be cost effective.
- Responses must be time sensitive.
- Responses must be realistic.
- Responses must have buy-in by the program team.

For risks that are a threat, there are three general categories of responses; avoidance, transference, and mitigation. *Avoidance* involves changing the program plan to eliminate the risk or condition or to protect the program's objectives from the impact. An example would be a decision not to use the latest or cutting-edge technologies, products, or procedures (sometimes I like to refer to these as "bleeding-edge"). Solely using BIM (building-information-modeling) for the development of the construction plans, use of a green roof system for LEED certification, or the use of an alternative project delivery method, are examples of what some might consider cutting-edge solutions they would like to avoid in their programs. *Transference* involves shifting the consequences of a risk to a third party along with ownership of the response. This generally involves payment of a risk premium to the party taking the risk. Examples include insurance, bonds, warranties, subcontracts, and so on. In *mitigation* the strategy is to reduce the probability or consequences of the adverse risk event to an acceptable level. Examples of mitigation strategies would be prequalifying specialty contractors, including an asphalt escalation clause in the construction contracts, or using bid alternates in an uncertain construction market.

For risks that are opportunities, there are three general responses as well; exploit, share, and enhance. The *exploit risk response* involves eliminating the uncertainty associated with the opportunity by taking actions that will ensure that the opportunity is realized. Examples would be helping a client develop a second project when the bids for the first project come in under budget or utilizing unique procedures, such as the bridging project delivery method, to improve the chances of success. The *shared risk response* opens up the benefits of the opportunity to a third party who would be more capable of making the positive event occur. An example is splitting savings on a contractor-generated value engineering idea. *Enhancing* an opportunity involves taking further actions to improve the probability or the impact of the positive risk event. An example would be to respecify full-depth paving instead of resurfacing, when the unit prices for asphalt are lower than expected.

In addition to planning for a specific response to a risk as outlined above, the program team may decide that the best solution is to accept the consequences of the potential outcome. This is done by either active or passive acceptance. Under active acceptance, the team identifies a risk and decides not to take a proactive action but instead establish a contingency plan. Contingency plans include setting of allowances or reserves to be utilized if the risk occurs. There are two types: a contingency reserve and a management reserve. *Contingency reserves* are for "known unknowns" such as providing a 1% allowance in the budget for bond premiums or including a two-week window in the program schedule for completion of punch list items. *Management reserves* are for "unknown unknowns" such as a 5% budget contingency for potential change orders or scheduling a program to end a month before the actual program deadline.

Finally, under passive acceptance, the program team identifies a risk but decides to deal with the risk if or when it occurs. This passive acceptance is for risks that are too small to be of concern. An example of such a strategy is deciding not to develop a risk response to the potential of a crash in the program management software. The likelihood of this event is small, and the cost to plan a detailed response to this risk is probably more than the cost to the program if it were to occur.

4.2.5 Developing the Safety Plan

The construction industry has always been a dangerous business. It is recorded, for example, that 30,609 people died building the Panama Canal between 1904 and 1914 [32]. According to the website of the US Bureau of Reclamation, 96 workers were killed during construction of the

Hoover Dam from 1931 to 1935. These deaths are classified as "industrial fatalities" from such causes as drowning, blasting, falling rocks or slides, falls, being struck by heavy equipment, truck accidents, and the like [33]. Although site safety has improved dramatically since these megaprograms were completed, in 2011 there were still an unacceptable 721 industrial fatalities at construction sites across the United States.

In 25 years in the construction business, personally I have experienced only one fatality . . . and will never forget it. The experience changed my life and approach to planning for safety. It was 1992, and happened during the first week of construction of a new highway system in Atlantic City, New Jersey. We were truncating an existing building to make way for an elevated section of the roadway. The building was located on private property, which we were renting as our construction field office. The first step in the demolition process was selectively to remove portions of the building's interior. One of the areas removed was a four-foot section of a loading dock. When it was removed, someone inexplicably placed a piece of three-quarter-inch plywood over the opening, presumably to allow for pedestrian access to the other side. Later that night a female employee of the property owner drove a forklift over the opening and crashed through the thin piece of plywood onto the concrete floor below. Along with a colleague, we discovered her lifeless body in the mangled forklift as we reported to work the next day. The gruesome scene cannot be described in words. And as the emergency crews were removing her body from the forklift, the sense of guilt overcame me. In our rush to start the project we had overlooked the most important thing, proper planning for site safety.

Developing a good safety plan involves three basic steps: establishing a programwide safety policy, determining safe work practices specific to each project site, and establishing an enforcement mechanism. The best place for the program team to start is the resources available through the Occupational Safety and Health Administration (OSHA, www.osha.gov). The US Congress, through the Occupational Safety and Health Act of 1970, established OSHA to insure safe and healthful working conditions for working men and women by setting and enforcing standards and by providing safety training, outreach, education, and assistance. The act covers private employers and their employees, including those in the construction industry, either directly through federal OSHA or through OSHA-approved state programs.

The first step in the planning process is agreeing on, and committing to, a programwide safety policy. The programwide safety policy must be established in recognition of the grave risks associated with human hazards on construction sites. OSHA reports that construction has the highest rate of accidents and fatalities than any other industry in the United States. The following is a list of the top-10 most frequently cited standards following inspections of construction worksites by OSHA [34]. OSHA publishes this list to alert employers about these commonly cited standards so they can take steps to find and fix recognized hazards addressed in these and other standards before OSHA representatives are on site.

- 1926.451–Scaffolding
- 1926.501–Fall Protection
- 1910.1200–Hazard Communication
- 1910.134–Respiratory Protection
- 1910.147–Lock-out/Tag-out
- 1910.305–Electrical, Wiring Methods
- 1910.178–Powered Industrial Trucks
- 1926.1053–Ladders
- 1910.303–Electrical, General Requirements
- 1910.212–Machine Guarding

The main purpose of the safety policy is to formulate the team's commitment to reduce the potential for these hazards and to ensure that the prevention of injury or illness will take priority over all other program goals. The safety policy is best established through brainstorming and must culminate in a written safety policy statement. The safety policy statement should be short and concise, easily understood, and agreeable to everyone on the program team. The following is an example of a good safety policy statement:

> It is the objective of the program team that all construction worksites are maintained in a safe, neat, and orderly condition, and free from human hazard. It is our policy that if an unsafe condition is encountered, that affected activity shall be suspended until the unsafe condition is corrected. The prevention of injury or illness will take priority over all other program goals.

Once the programwide safety policy is established and committed to, site-specific safety plans must be developed for each construction project.

The OSHA mandates and delineates the contents and requirements for these site-specific plans. In general the requirements include:

- *Safety Rules.* The site-specific safety plan must outline the safety rules as they apply to each construction project. These rules should be short, concise, simple, and enforceable, and at a minimum cross-reference the applicable OSHA statutes.
- *Safety Responsibility.* The entity, and critically the individual, with the responsibility and authority to enforce safety requirements must be clearly defined in the site-specific safety plan. Even though well-defined legal, contractual, or insurance conditions may have predetermined specific safety roles and responsibilities, they nonetheless need to be summarized and formalized. The following is a good example of a written statement regarding safety responsibility:

  > The Contractor has the sole responsibility for ensuring that the construction worksite is safe, neat, and maintained in an orderly condition and is free from safety and health hazards. The Contractor is also solely responsible by law for compliance, and regulatory reporting requirements, for all workplace and employee safety and health issues. The Contractor's designated On-Site Safety Representative will be the sole point of contact for all safety issues and shall have the authority to stop work and implement corrective procedures.

- *Site Access and Control.* As the example of the fatality from the Atlantic City program illustrates, the access to a construction site from outside sources must be continuously monitored and controlled. Maintaining construction site security is a critical component of site safety and must be properly planned and managed. This is dependent on the nature and location of the particular site; that is, an urban area will require different security measures than a rural area.
- *Worksite Analysis.* The site-specific safety plan should require that an analysis be routinely conducted of all areas of the construction site by the On-Site Safety Representative. After each inspection, a written report should be completed and retained for the record. The report is designed to address any unsafe conditions or unsafe acts. The report will also include what corrective actions are to be taken and who is responsible, and accountable, for the correction of defects.

- *Safety Training.* The safety plan will require that the contractors implement a training program that provides orientation and training for each new construction worker, or when new equipment processes or procedures are initiated. The training will consist of, but not be limited to, correct procedures to follow, correct use of required personal protective equipment, and where to get assistance when needed. In addition, weekly "toolbox meetings" will be mandated that will cover general safety topics for all construction workers. Records must be maintained showing the safety topics discussed and names of those attending. Safety meeting topics should be designed to instruct workers on how to perform their jobs productively, efficiently, and safely. In addition, recent work area inspection results, workers' compliance with safety procedures, and the accident investigations that occurred since the last safety meeting will be reported. The safety plan will also require that training be provided to all persons in construction supervisory positions. This training will consist of, but not be limited to, correct procedures for conducting safety meetings, conducting safety inspections, accident investigation, job planning, employee training methods, and task analysis.
- *Accident/Incident Investigation Procedures.* The safety plan should require the construction contractors to follow all OSHA requirements regarding reporting of accidents or injuries. The accident investigation report must include information required to determine the basic causes of the accident and what corrective action is to be taken and/or recommended to prevent a recurrence of a similar accident
- *Record Keeping.* The construction contractors are required by OSHA to keep records of work-related fatalities, injuries, and illnesses. In addition to these OSHA logs, which are retained for five years (a federal requirement), each construction contractor will be required to maintain other safety records for a period of one year from final acceptance of the construction work. These will include inspection reports, accident investigation reports, minutes of toolbox safety meetings, and training records.
- *Emergency Action Plan.* The program management team will develop a written emergency action plan to ensure to the extent possible the safety of all employees, visitors, contractors, and vendors at each construction site at the time of emergency situations, such as but not limited to natural disasters, fire, explosions, chemical

spills and/or releases, and medical emergencies. Evacuation routes will be required to be posted in all work areas showing primary and secondary routes for employees' evacuation to a safe, predetermined location for a head count. Contact information, and travel routes, to the nearest fire station, police station, and health care facility will also be required to be posted at the site.

The last part of the safety planning process is to formalize the enforcement mechanisms that will hold the team accountable for the written safety procedures. OSHA rules and mandates will dictate the enforcement mechanisms and penalties for noncompliance for programs in the United States. Although I focused on the United States OSHA program in this section, there are similar governmental agencies in other countries with a similar mission and workplace safety rules, including:

- Canada: The Canadian Centre for Occupational Health and Safety (CCOHS) is a Canadian federal government agency whose mission is the elimination of all Canadian work-related illnesses and injuries. CCOHS has workplace safety rules and regulations similar to OSHA.
- European Union: The European Agency for Safety and Health at Work (EU-OSHA) mission is making Europe a safer, healthier, and more productive place to work. They promote a culture of risk prevention to improve working conditions.
- Peoples Republic of China: In China the Ministry of Health is responsible for occupational disease prevention and the State Administration of Work Safety for safety issues at work. On the provincial and municipal level, there are Health Supervisions for occupational health and local Bureaus of Work Safety for safety.

4.2.6 Developing the Change Management Plan

My first project out of graduate school in 1989 was the construction of the Beth Israel Garage in Boston, Massachusetts. The new garage was a cast-in-place concrete, five-level, below-grade parking facility for 750 vehicles. We utilized the then state-of-the-art technique of up-down construction with slurry walls to build the garage. Our work also included the demolition of the 200-year-old Massachusetts College of Arts building, with preservation of its historic façade, and a separate contractor was responsible for construction of a new 12-story, 380,000 sf Clinical Health

FIGURE 4.10
Beth Israel Clinical Health Center.

Center in its place. The total cost of the program exceeded $100 million. Construction was technically challenging as while we were mining and placing concrete for the underground garage, working from the surface down, the new clinical center was going up directly over us (see Figure 4.10).

Although remarkable for many reasons, the thing I remember most about the project was the first construction coordination meeting with the program manager. The program manager was a distinguished-looking, seasoned architect, probably in his 60s at the time. He was well-spoken, commanding, and as a young man new to the profession, he garnered my immediate attention and curiosity. About halfway through the meeting, he confidently declared, "This program will have no change orders." Needless to say, he lost a lot of his credibility with the arrogance of that statement. Publius Syrus, a Latin writer of maxims in the first century BC, put it best, "It is a bad plan that admits of no modification." All construction programs will have changes and therefore we need to plan for them. Depending on the nature of the program, change orders, on average, will account for between 2.5%–7.5% of total program cost.

Change orders can be grouped into three general categories: unforeseen conditions, design issues, and changes in scope. An unforeseen condition is an unanticipated or unexpected circumstance or situation. Typical unforeseen conditions on construction sites that have the potential to affect the program's schedule or budget significantly include:

- Differing subsurface conditions (i.e., discovering rock during excavation for a foundation)
- Surprises uncovered during renovation of an existing facility (e.g., asbestos insulation on mechanical piping)
- Severe weather
- Changes in building codes (enacted after the program starts)
- Labor unrest

The next category of change orders is design issues, which generally entail errors or omissions on the plans and specifications. These are often the most controversial and misunderstood category of program changes. Architects and engineers do have a professional obligation to design in accordance with a reasonable standard of care. That does not mean that the design will be perfect, inasmuch as it is unreasonable to believe that the complex process of creating the plans and technical specifications can be done without error. As a result, design-related changes are inevitable on any construction program. Yet the owner implicitly warrants the adequacy of the plans and specifications during the bidding process. This has been tested in the courts and is referred to as the Spearin Doctrine [35]. So design issues are bound to happen and without doubt will result in change orders. The program team must accept and plan for them.

Changes in scope make up the third category. Scope changes should always be dictated by the owner. This is the case because, depending on the nature of the change in scope, it may require a revisit, and potential modification, to the program management plan. Common types of scope changes include design alterations, quantity changes, and schedule modifications. Scope creep, defined as the unintended increase in scope, is unacceptable and must be avoided. The best way to eliminate scope creep is to properly monitor and control the design process. We cover more on that later in Chapter 5.

The change management plan sets the rules of engagement for managing each category of change request, on both the program and project levels. The change management plan provides a template for managing change by developing procedures to:

- Identify the type of request for change.
- Determine if the change affects the program (scope, budget, schedule).

- Control the way that change is undertaken.
- Manage the approval of change.

Many standardized change processes have been developed for construction programs. The particular change process will be dependent on the type of program, whether it is private or public work, and the type of delivery method. A general flowchart for change management is provided in Figure 4.11. It is critical that the program team agree on a process for change management including delineating the levels of authority for approvals and sign-offs. The process should be streamlined as much as

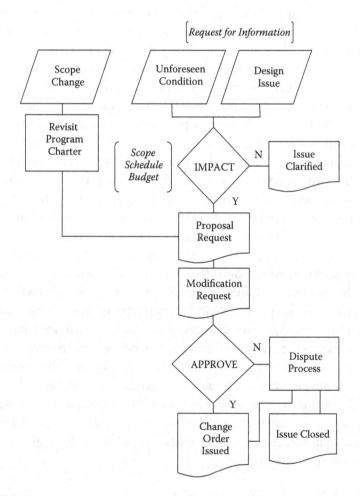

FIGURE 4.11
Basic change management flowchart.

possible so that approvals do not result in project delays or unnecessary financial hardship on the contractor.

4.2.7 Developing the Communications Management Plan

Proper communication is the key to implementing plans and strategies, and ultimately to delivering the program benefits. The distribution of information in a timely and accurate fashion is essential for success. But as aptly put by George Bernard Shaw, "The single biggest problem in communication is the illusion that it has taken place." A good communication management plan will make sure the right level of effective communication is taking place both internally and externally

On the program level, communication is critical as it enhances collaboration. Effective teams are characterized by trust, respect, and collaboration. This is particularly essential during the construction initiation and planning processes, where it is estimated that over 75% of all program decisions are made [36]. An effective communications management plan will allow for an open, but controlled, flow of information. This will facilitate teamwork so there are no surprises. It is also critical that a governance mechanism, which will establish communication procedures when issues escalate, is set and agreed upon by the team. To make that happen, the team must first agree on a communications strategy. This is best described as determining "How much information will be sent to whom, how, and how often" [6]. The strategy will be different for internal versus external stakeholders.

For construction programs, internal stakeholder communications are driven by the project delivery method(s), which sets the roles and responsibilities of the team. Internal communications will focus on team collaboration and accountability, referred to in the trade as setting the "ball in court." This term comes from tennis, where it means it is the opponent's turn to serve. In other words, it sets who is responsible for what and when. Many excellent comprehensive communications management systems have been developed around this concept for the standard tasks that make up a typical construction program. These include web-based computer software programs such as Primavera Contract Manager®, Prolog®, and Constructware®. These systems will establish the program communication channels based on the project delivery method(s), set ball in court metrics including task durations and required-by dates, and regulate rule-based alerts for deficiencies or missed milestones.

It is also critical to have face-to-face interchanges with the program team to build community and relationships. A series of program meetings must be agreed on, and formalized in the communications management plan. Sometimes we have fun with this concept, explaining there is a need to have a meeting to plan for meetings. A standard meeting schedule that has worked successfully for my construction programs is provided in Table 4.4. If a significant change occurs, a special meeting may need to be called specifically to address the issue at hand.

External communications will focus on accountability and transparency. Transparency requires that the decisions and actions of the team are open to scrutiny and acknowledgment that the program's external stakeholders have a right to access such information. This is critical, as the external stakeholders must develop respect for the team and have trust, throughout the program life cycle, that the program will realize its anticipated benefits. The communications management plan should include process and procedures for encouraging open and honest relationships with

TABLE 4.4

Meeting Schedule

Meeting	Attendees	Frequency	Distribution	General Purpose
Owner	Owner Program Manager Design Professionals Construction Professionals	Biweekly	Attendees Internal Stakeholders	Program Issue Resolution
Progress and Coordination	Owner Program Manager Project Managers Design Professionals Construction Professionals	Weekly	Attendees External Stakeholders Internal Stakeholders	Progress Update Program Coordination
Superintendent	Program Manager Project Manager Design Professionals Construction Professionals Contractors	Biweekly	Attendees	Project Issue Resolution Coordination

external stakeholders, including the general public, if applicable. Tools to engage with stakeholders include:

- *Program Status Reports:* Program status reports, prepared and distributed on a minimal monthly basis, are essential tools for communicating with stakeholders. A well-written status report will be succinct yet detailed and will inform stakeholders what has been done, what needs to be done, what is being waited for, and any problems and proposed methods to resolve them.
- *Meetings with Stakeholders:* Starting at the beginning of planning, and commencing at program closure, regular meetings should be held with interested and influential stakeholders to update them personally of the status of the work. This will also give the program team time to interact with the critical stakeholders and more fully understand and appreciate their issues. If a significant change occurs, a special meeting may need to be called specifically to address the issue at hand.
- *Program Website:* A great way to enhance accountability and transparency is through a program website. This is especially important on public programs where there may be a greater potential for construction to affect peoples' daily lives. The website should focus on the positive aspects of the program, provide periodic progress updates, and provide construction alerts for things such as traffic delays or other environmental impacts such as increased dust or noise. For full transparency, a webcam(s), with a view(s) overlooking the entire construction site(s) can be provided.

4.2.8 Establishing Rules of Engagement

A critical part of the program management plan is the establishment of standard guidelines for how the team will work effectively together to implement the program charter. Construction is unique from other programs in that a major contributor to the team, the construction contractor, is often not at the table during the creation of the program charter. For most public projects, under the mandated design–bid–build project delivery method, selection of the contractor does not take place until the end of the planning phase. At this point the program charter has been even further developed into the detailed strategies and procedures of the program management plan.

One of my colleagues used to point out to me that one of the major differences between "family" and "team" is that you get to choose who's on your team. In construction programs the initial team members are generally chosen through a qualifications-based selection process. For programs that expend public monies, United States federal and state statutes permit this type of procurement for technical consultants, such as the program manager, financial advisor, legal team, and the design professionals. Under this approach, selection is based on credentials, reputation, recommendations, and more subjective factors such as the potential to work effectively with the team and alignment with the program's core values. On public programs, this is often not the case for the construction contractors. For them, the project sponsor is often required by law to choose the lowest responsible, responsive bidder. A *responsive bidder* generally means a contractor who has submitted a bid that conforms, in all material aspects, to the procurement documents. A *responsible bidder* refers to a contractor who has the capability to perform per the contract requirements and has the moral and business integrity and reliability that will ensure a good faith effort. The low-bid selection process is done with the intent of openness and fairness to give equal opportunity to public monies. A flaw in this type of procurement is that it assumes that all firms will bid the work at their lowest possible cost based on a reasonable interpretation of the bid documents and will proceed with the work in good faith. Unfortunately, my experience is that this is not always the case. So because the low bid firm is not really "chosen" to be part of the team, analogously this can lead to the selection of "Uncle Henry" as the contractor . . . that guy at the family picnic who disrupts the whole affair. In other words, like with your family, you have to make do with who you have, and this can lead to unfortunate results if not managed correctly.

As discussed in Chapter 2, one way of dealing with this situation is to have a tight set of general requirements and a coordinated set of technical plans and drawings. They should be clear and concise, establish the program goals, set the general rules of engagement for the team, and provide a detailed path to the desired outcome. But even under the best of conditions, the bidding documents are only representational and therefore require interpretation by the contractors. To increase the likelihood of success, additional program-specific contract provisions should be developed during the planning phase for inclusion in the procurement documents. They will focus on integrating the contractor with the rest of the program team and further define and communicate the rules of engagement already established in the program charter. Critical items to address include defining program roles and level of authority, setting planning and scheduling responsibilities for the team focusing on integration of the contractor's work schedule with the master program schedule, defining and setting important intermediate contract milestones, setting policies and procedures for worksite safety, and defining the dispute resolution process.

I refer to these as "special project provisions," inasmuch as they are specific (special) to the program. Below are examples of special project provisions that I have put together for these critical program issues. I have abridged them here to more clearly demonstrate their intent and to remove some of the legalese that is an unfortunate necessity in all contracts [37].

4.2.8.1 Roles and Responsibilities of the Program Team

The intent of this provision is to help the contractor understand the relationship and organizational hierarchy of the main participants on the program team and how that may affect its work. Because the contractor most likely did not have the advantage of participating in the chartering sessions where these early decisions were made, it is critical during the bidding process to share this information with them so they know what they are getting into. An example from one of my recent programs:

A Program Manager has been retained for the purpose of assisting the Architect in administration and coordination of the contract. The Program Manager will approve the Contractor's proposed construction schedule and observe the Contractor's rate of progress. It is the Contractor's sole responsibility to monitor the rate of progress of the work to ensure that the project is completed within the time frame stipulated in the Contract Documents and that each scheduled project milestone is met. However, if in the view of the program manager, the contractor is in jeopardy of not completing the work on time, or not meeting any schedule project milestone, the program manager may request that the contractor submit a recovery schedule. The recovery schedule shall show, in such detail as is acceptable to the Program Manager, the Contractor's plan to meet all scheduled project milestones and that the work will be completed within the time frame stipulated in the Contract Documents. The Architect and Program Manager shall jointly observe the work to determine when conditions precedent to Substantial Completion and Final Acceptance [have] been fulfilled by the Contractor. Upon acceptance of the work by the Architect, and acceptance by the Architect and Program Manager that all conditions of the Contract Documents have been fulfilled by the Contractor, notice of Substantial Completion and Final Acceptance shall be granted to the Contractor by the Architect.

4.2.8.2 Planning and Scheduling Requirements

The intent of this provision is to integrate the planning and scheduling concepts that the contractor used as a basis for their bid with the master program schedule developed by the team during the chartering sessions. It also sets the rules of engagement for the transfer of planning, schedule, and progress data from the contractor's team to the program team. The current provision is detailed and comprehensive, which I have learned is a necessity, based on lessons learned from past missteps and mistakes. This is further explored later in the case study at the end of the chapter. The provision includes procedures for developing and updating the master program schedule based on the contractor's input and reads in part:

> The Program Manager will prepare a coordinated computerized Original Baseline Schedule based on the schedule input of the Contractor. The Contractor shall designate a representative to be responsible for the CPM scheduling functions relative to their contract and such person shall be the liaison between Program Manager and the Contractor. The Contractor shall provide a list of work activities that shall be a comprehensive inventory of all work activities that will comprise the Contractor's work on the project. This shall include all contract times, the project start and finish dates, submission and approval of all project deliverables, all required tasks in the procurement cycle (submission of submittals, approvals, fabrication, delivery to the site), all construction work tasks, project closeout tasks including punch list and equipment testing, and all tasks required for the final acceptance of the work. The Contractor shall also be responsible for scheduling of subcontractors and suppliers. In preparing the List of Activities, the Contractor will be required to furnish a brief description of each activity, durations, predecessor and/or successor activity(s), phase codes, area codes, responsibility codes, revenue/cost loading discretely by task and equipment and person-hour requirements discretely by task. After receiving the initial scheduling information from

the Contractor, the Program Manager will develop a Preliminary CPM Schedule incorporating the schedule data provided. Any activity float time will be included in the Preliminary CPM Schedule at the discretion of the Program Manager. The Preliminary CPM Schedule will be presented and discussed at a Scheduling Meeting, called by the Program Manager and attended by the Contractor. At this meeting, the Program Manager will explain the Preliminary CPM Schedule in detail. During the presentation, the Contractor shall indicate their views, their approval or shall request changes. The Program Manager will make all changes to the Preliminary CPM Schedule that are generally compatible with the proposed activities and requirements of the Contract Documents and which have been agreed to previously by the Contractor. After the Scheduling Meeting, the Program Manager will produce an Integrated Baseline Schedule. One copy of the Integrated Baseline Schedule will be provided to the Contractor. The Contractor will sign the original of the network diagram indicating their approval of the Integrated Baseline Schedule. It will be the responsibility of the Contractor to insure that all of their work is incorporated into the Integrated Baseline Schedule and that it correctly represents the means, methods, techniques, sequence and procedures in which they plan to complete the work. Once the Integrated Baseline Schedule is approved by the Contractor, it will be used as the basis to monitor schedule progress. At the end of each calendar month, the Contractor will review the Integrated Baseline Schedule with the Program Manager. Prior to this meeting, the Contractor shall prepare a typewritten Activity Status Report detailing each activity in progress, giving percentage completed, remaining duration, summary of delays in starting or finishing an activity, etc. In this report, the Contractor shall also indicate what steps are being taken to correct delaying conditions. Based on this information, the Program Manager will prepare an update to the Integrated Baseline Schedule. In the event that the updated Integrated Baseline Schedule indicates that the Contractor has been delayed in execution of their work, and that this has impacted the critical path, the Contractor may either request an Extension-of-Time or will be required to recover the lost time. Any request for an Extension-of-Time must contain a CPM type schedule analysis, performed by the Contractor, which shows, in a level of detail that is satisfactory to the Program Manager, the impact of the delay to the critical path and the project milestones. Based on this analysis, the Program Manager may either grant the Extension-of-Time or require a Recovery Plan. If requested, the Recovery Plan shall show, in such detail as is acceptable to the Program Manager, the Contractor's plan to meet all schedule project milestones and that all work will be completed within the time frame stipulated in the Contract Documents. Explanations for schedule recovery may include items such as adding additional resources to accelerate activities on the critical path, working additional hours, working through holidays and weekends, change in means and methods or revision of the overall sequence logic of the Integrated Baseline Schedule to adjust the critical path.

4.2.8.3 Critical Intermediate Construction Milestones

In this provision the critical intermediate construction milestones are defined and set. These milestones are extracted from the conceptual construction schedule developed by the program team during creation of the detailed master schedule. An example from my experience as a contractor in upstate New York in the United States, would be requiring a facility to be "weather-tight" prior to the onset of the winter months. This is done to ensure that the interior work can be accomplished without interruption during the harsh conditions of winter and spring. Consideration should be given to the type and number of intermediate construction milestones set as contract requirements. Too many could imprudently influence the contractor's means and methods, and too few could jeopardize the ability of the program team to properly monitor progress.

4.2.8.4 Critical Contract Milestones

In this provision the critical contract milestones are defined and set. A contract milestone must be clear and unambiguous. A contract milestone must also have a single owner who is solely accountable for achieving it. Contract milestones are different from intermediate construction milestones as they have predetermined consequences if they are not met. For construction programs, especially those with liquidated damages or penalties for delay, the definition of the completion milestones is most critical. There are generally three critical completion milestones: beneficial occupancy, substantial completion, and final acceptance. Beneficial occupancy is the use or occupancy of the work by the owner, even though all of the contractor's work is not yet substantially complete. Substantial completion is the declaration by the contractor that the work is finished except for minor items that need to be corrected or completed as detailed on a punch list. The punch list is a list of tasks or "to-do" items. The phrase takes its name from the historical process of punching a hole in the margin of the document, next to the completed items on the list. Final acceptance is when the punch list is complete, and other contract requirements for closure are met. Similar to the scheduling provisions, through lessons learned, these provisions have become quite comprehensive:

The Owner reserves the right to use or occupy all or parts of the work at the Owner's sole discretion, and before the work or part thereof is substantially complete. However, unless specifically scheduled otherwise, or by prior agreement, the Owner shall not be required to use or occupy the work or any part thereof until all of it is substantially complete. Beneficial occupancy of the work or part thereof by the Owner shall not relieve the Contractor from completing all the work in accordance with the Contract Documents or from other contractual obligations, and shall not prejudice the Owner in any way. In the event the Owner takes Beneficial Occupancy of the work or designated part thereof, the Architect shall prepare and issue to the Contractor a Notice of Beneficial Occupancy, clearly identifying the occupied work, the Contract value of the occupied work, the date of Beneficial Occupancy, the beginning and end dates of the guarantee period of the occupied work and the continuing responsibilities of the Owner and Contractor for operation, maintenance, utilities, security, insurance, etc. Generally, but not necessarily, the guarantee period for occupied work will not commence until the work is substantially complete, as hereinafter described. Generally, but not necessarily, the retainage amount associated with the occupied space will not be reduced until the work is substantially complete, as hereinafter described. Substantial Completion: When the Contractor has completed the work, or designated parts thereof, to a point that, in the opinion of the Contractor, the work is substantially complete, the Contractor shall so notify the Architect in writing. However, unless specifically scheduled or agreed to in advance by the Owner, the Owner shall not be obligated to consider any part of the work for substantial completion until all of the work of the contract is substantially complete. As soon as reasonably practical after receiving such notification, the Architect and Program Manager will inspect the work, and thereafter, advise the Contractor of any deficiencies or other impediments to determining the work to be substantially complete. Note that any such inspection and listing of impediments to substantial completion shall not be construed to be a "final inspection" or "punch list," unless specifically identified as such by the Architect. When the Architect and Program Manager determine that the work is, in fact, substantially complete, a final inspection involving all interested parties will be scheduled and conducted by the Architect and Program Manager. The Owner's operation and maintenance personnel may participate in this inspection or may perform their inspections separately. Following the inspection(s), the Architect will provide the Contractor with a compiled list of defective, deficient, incomplete or otherwise unacceptable work. This list is commonly referred to as a "punch list." The Architect will indicate on the punch list his opinion of the estimated cost of completing or correcting each of the items listed thereon. After preparation of the punch list, the Architect will prepare and issue a Certificate of Substantial Completion. This document will clearly identify the parts of the work which are substantially complete, the value of the substantially completed work, including any fully executed change orders applicable thereto, the date of substantial completion, the beginning and end date of the guarantee period, and the continuing responsibilities of the parties for operation, maintenance, utilities, security, insurance, etc. The punch list will be attached to the Certificate of Substantial Completion and be made a part thereof. (The value of substantially completed work shall be determined from the bid items, or, if no applicable bid items exist, from the Contractor's approved lump sum breakdown.) After all of the remaining items of work indicated on the punch list(s) are satisfactorily completed or corrected, the Owner will release the amount withheld for these items, upon submission of affidavits showing that all claims, liens, judgments, laborers, vendors and sub-contractors have been paid in full or otherwise discharged. Partial releases of monies retained for punch list items will not be made. In the event the Contractor fails or refuses to satisfactorily complete or correct the remaining items of work within sixty (60) calendar days from the date of Final Payment, the Owner reserves the right to have the work completed or corrected by others and to deduct the cost thereof from monies otherwise due the Contractor.

4.2.8.5 Safety Policies and Procedures

The owner transfers the risk of construction site safety to the contractor through the execution of the contracts. It is therefore imperative that the contracts contain proper language detailing the specific nature of this significant risk transfer. The program safety plan will be used as the basis for developing this provision. Following is a good example of a well-prepared, comprehensive, safety statute from a recent program:

> It is the objective of the Owner that the Contractor maintains the construction worksite in a safe, neat, and orderly condition, and free from human hazard. It is the policy of the Owner that if an unsafe condition is encountered, that affected activity shall be suspended until the unsafe condition is corrected. It is the policy of the Owner that if an unsafe condition is encountered, that affected activity shall be suspended until the unsafe condition is corrected. The Contractor has the sole responsibility for ensuring that the construction worksite is safe, neat and maintained

in an orderly condition, and is free from safety and health hazards. The Contractor is also solely responsible by law for compliance, and regulatory reporting requirements, for all workplace and employee safety and health issues. The Contractor's designated On-Site Safety Representative will be the sole point-of-contact for all safety issues and shall have the authority to stop work and implement corrective procedures. The Contractor is required to submit a Project Specific Safety Plan. The Project Specific Safety Plan shall outline the Contractor's actions that will ensure that the Project site is maintained in a safe, neat and orderly condition, and is free from recognized hazards that could cause injury or death. The Project Specific Safety Plan shall also contain written procedures for the Contractor's compliance with governmental safety laws and associated reporting requirements. At a minimum, the Project Specific Safety Plan shall include each of the following:

- Project Safety Objective Statement
  - Safety Responsibilities and Roles within the Contractor's Organization
  - The Contractor's Safety Policy Requirements for Sub-contractor, or a written agreement to follow the Contractor's Safety Policy
  - Mandatory Guideline for the use of Personal Protective Equipment
  - Emergency Response Procedures, including Routes of Egress and Assembly Areas
  - Procedures for Investigating and Reporting Accidents
  - Site Security Procedures
  - Procedures for Governmental Agency Compliance Reporting
  - Procedures for the Protection of Project Site Visitors
  - Safety Procedures related to the Maintenance and Protection of Traffic, including flagging operations
  - Hazard Communication Program–Location of MSDS [38]
  - Lockout/tag out and Ground Fault Protection Procedures
  - Identification of the proposed Site Safety Representative and Competent Person, including credentials
  - Hazard Analysis for all Major Work Areas
  - Rigging and Crane Safety Procedures, including required Inspections
  - Statement acknowledging that the Contractor is solely responsible for construction worksite safety issues
- Statement on Excavation/Trenching Responsibility
- Identify who is responsible for
  - First aid/equipment and supplies, may include eye wash stations if corrosives are in use on site

- Fire protection and Fire prevention (Fire extinguishers and training)
- Housekeeping requirements (removal of debris)
- Field Sanitation – Toilets, supplier of potable water
- For Steel Erection – (there must be a separate site safety plan):
  - Identifies order of erection
  - Laydown and staging areas
  - Site roadways
  - Plan to prevent elevated loads above other activities
  - A steel erection Contractor shall not erect steel unless it has received written notification that the concrete in the footings, piers and walls or the mortar in the masonry piers and walls has attained, on the basis of an appropriate ASTM [39] standard test method of field cured samples, either 75% of the intended minimum compressive design strength or sufficient strength to support the loads imposed during steel erection
- Identify lines of responsibility for fall protection/prevention at all stages of construction.
- Contractor and sub-contractors shall provide competent persons for the erection and dismantling of scaffolds.

4.2.8.6 Alternative Dispute Resolution

Alternative dispute resolution (ADR) is a general term encompassing various techniques for resolving conflict outside the legal system. A number of different ADR methods are currently used in the construction industry. During the planning process the program team must agree on whether there is a strategic advantage to leverage ADR, and if so, what method(s) to utilize. Some of the more common methods are:

- *Step Negotiation:* This ADR technique requires the entities directly involved in the dispute to seek resolution through one-on-one negotiation first. If a resolution is not reached within a predetermined length of time, the dispute is elevated to the next level in the organization. During planning, a committee is set up at each level of organizational hierarchy and is given the authority to solve issues through compromise or conciliation.
- *Dispute Review Boards:* This typically consists of three neutral experts, who visit the site periodically in order to monitor progress and potential problems. When requested by the parties, the board conducts an informal hearing of the dispute and issues

an advisory opinion that the parties use as a basis for further negotiations.

- *Nonbinding Mediation:* This is an ADR process where an impartial person, the mediator, facilitates communication between parties to promote reconciliation, settlement, or understanding among them. The mediator helps the parties identify the important issues in the dispute and decide how they can resolve it themselves.
- *Binding Arbitration:* This is a forum in which each entity and counsel for the entity presents the position of the party before an impartial third party, who renders a specific award. It is agreed beforehand that both parties will accept the award bestowed by the arbitrator.

There has been a trend in the construction industry toward ADR as a way of reducing the high degree of unresolved disputes that require settlement in the court system. The most recent version of The American Institute of Architects (AIA) standard forms added a mediation requirement prior to binding arbitration for all disputes. The Engineers Joint Contract Documents Committee (EJCDC) standard forms require good faith negotiations for 30 days and allow for more definition of dispute resolution, including options for requiring either mediation or arbitration. The Design–Build Institute of America (DBIA) standard forms require step negotiation, followed by mediation, prior to binding arbitration for all disputes. The most successful form of ADR in my view is step negotiation. We used this form of ADR on the $70 million Clinton St. Storage Project, where we successfully implemented the step process shown in Figure 4.12 for all issue resolutions.

Teams were assigned the responsibility and authority to resolve issues at each level and each team's leader was given the additional role to determine if an issue needed to be moved up to the next level. The process worked very well with most issues being resolved at the filed level (Steps 1–3). There was one complex issue, however, that was delegated all the way up to the Executive Office level.

Shortly after winning the low bid contract, the tunneling contractor (my old colleagues from Kiewit) proposed a radically different approach to the project from what we anticipated during the planning phase and presented in the bid documents. The Clinton St. Storage Facility is a 6-million-gallon underground CSO located in downtown Syracuse, New York. The facility was designed as three parallel 18-foot diameter, underground

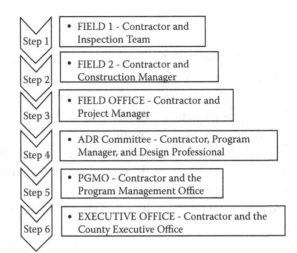

FIGURE 4.12
ADR step negotiation process.

FIGURE 4.13
Cut and cover approach (Clinton Storage Facility).

storage tunnels. Similar to the other LIPO CSO projects, wastewater was designed to be stored in the three 850-foot-long tunnels until it could be conveyed to Metro for treatment. Kiewit proposed a "cut and cover" approach (see Figure 4.13) instead of shielded soft-ground tunneling using a tunnel boring machine (TBM). The shielded soft-ground tunneling method was envisioned by the program team because the soil conditions at the site (clay, silt, and sand) would require this type of tunneling if a TBM were used. Cave-ins are a constant threat when tunneling in soft ground. To prevent this from happening, the design professionals specified a special piece of equipment called a shield. A shield is an iron or steel cylinder literally pushed into the soft soil. It carves a perfectly round hole and supports the surrounding earth while the TBM removes debris and installs a permanent lining made of cast iron or precast concrete. When the workers complete a section, jacks push the shield forward and they repeat the process. Although technically a correct solution to the design challenge, it was later learned that this was not the right approach from a cost and schedule perspective (Figure 4.13). The Kiewit team realized this and to increase their chances of being the low bidder priced the cut and cover method instead. The cut and cover method allowed them to support the excavation with conventional methods (steel sheeting and slurry walls) and construct the tunnel from the surface. It is estimated that this approach alone saved $14 million and most likely was the only way to get the project done on time.

Even though this was clearly the better approach, because of the timing, it would in essence require a change in the project delivery method from design–bid–build to design–build. It would also require significant changes to the contract, the cost and effort of redesign, and a revisit of life-cycle issues such as operational and maintenance requirements. Also problematic was insuring the integrity of the bidding process, as from what we could determine; all of the other bidders based their cost on the soft tunneling method.

Although the team agreed this was the best approach these concerns had to be worked out before Kiewit's plan could be implemented. The step process was very useful as the concerns regarding implementation were resolved by the team at each appropriate ADR level and the ultimate decision to proceed was properly delegated to the execution office. The process expedited the decision which ensured that the time and effort for considering Kiewit's approach would not negatively affect the project or program. In the end Kiewit's "value engineering" proposal was accepted and a credit change order was negotiated and executed to everyone's satisfaction.

In order to implement an ADR process such as the one described above, the program team must first develop an outline of the expectations and responsibilities of each party in the process. As the contractor will be an integral part of the ADR process, a special provision, such as the example below, must be included in the construction contracts.

Any controversy or claim arising out of or related to the Contract, or the breach thereof, shall be settled by mediation as outlined herein, unless the parties mutually agree otherwise. Such controversies or claims upon which the Architect has given notice and rendered a final decision shall be subject to mediation procedures as outlined below upon written demand of either party. The Owner and Contractor will attempt in good faith to resolve any controversy or claim arising out of or relating to the Contract, its breach, termination, or validity through non-binding mediation. Mediation proceedings shall take place at a location in (place of jurisdiction) that is mutually acceptable to both parties. The Owner and Contractor will mutually agree on the designation of a neutral third party who will act as the mediator during the dispute resolution process. The role of mediator is to guide the dispute resolution process, insuring that each party has an uninterrupted opportunity to speak and respond, until either resolution or stalemate. If the controversy or claim has not been resolved pursuant to the mediation procedure within 75 days of the commencement of such procedure, then the controversy shall be settled either by:

- Arbitration in accordance with the Construction Industry Arbitration Rules of the American Arbitration Association currently in effect or such other rules as the parties mutually agree upon, provided that the parties mutually agree and consent to arbitration: or
- If the parties reject arbitration, then they resort to litigation proceedings in (place of jurisdiction) which court shall have exclusive jurisdiction.

Demands for mediation shall be filed in writing with the Owner. A demand for mediation shall be made within 21 days of the Architect's final decision. In no event shall the demand for mediation be made after the date when institution of legal or equitable proceedings based on such claim, dispute, or other matter in question would be barred by applicable statutes of limitation. No mediation arising out of or relation to the Contract shall include, by consolidation, joinder, or in any other manner, an additional person or entity who is not a party to the Contract, except by written consent containing a specific reference to the Contract signed by the Owner and Contractor, and any other person or entity sought to be joined. Consent to mediation involving an additional person or entity shall not constitute consent to mediation of any claim, dispute, or other matter in question not described in the written consent or with a person or entity not named or described therein.

4.2.9 Developing the Transition Plan

The transition plan establishes processes and procedures for transitioning the program from accomplishment of the desired outputs and outcomes to the attainment of its benefits. It will ensure that the program, and the constitute projects, are taken to closure and then operational status in an efficient and responsible manner. It will also ensure that benefits are sustained throughout the useful life of the program.

To achieve this, the transition plan will outline the quality assurance-based processes for delivery, verification, and managing risks to critical functions performed in, or by, the projects constructed as part of the program. This is done by making sure the plans and specifications are implemented as intended, and the completed projects are functioning correctly. The transition plan will also ensure that the program will deliver construction projects that meet the end user's needs, at the time of closure. The transition plan will include, as a minimum, the following elements:

- Clear organizational requirements for transitioning projects from execution to closure and then to operation, including level of authority for sign-offs
- Guidance to integrate life-cycle considerations, such as operational efficiency and ease of maintenance into planning
- Processes and procedures to ensure that operation and maintenance personnel are properly trained
- Standards of performance and processes to verify that the projects meet those standards, including start-up and testing procedures for critical components
- Procedures to prepare and document a coordinated set of project record drawings, sometimes referred to in the trade as "as-builts"
- Procedures to prepare and document a comprehensive set of operation and maintenance manuals
- Requirements for warranties

In addition to the constitute projects, the program itself will need to be transitioned. Once all projects are closed and transitioned properly to operational status, and the benefits of the program are achieved, the program will then transition to formal closure. We discuss the program closure phase in detail in Chapter 6.

4.2.10 Case Study of an Effective Planning Process

An effective planning process is a prerequisite to the successful execution of a program. During the construction phase, where the majority of progress monitoring and controlling takes place, decisions made during planning will give the program team the tools they need to ensure on-schedule, within scope, and under-budget, completion. In construction, the constraints of schedule and budget are considered the most critical elements to control and monitor. In fact, an owner's decision whether to engage a construction professional during the planning process is often determined based on the importance of these two elements to the program's success. That was certainly the case with the Atlantic City/Brigantine Connector program.

Construction of the $330 million Atlantic City/Brigantine Connector was the largest design/build program ever undertaken by the State of New Jersey. Linking the Atlantic City Expressway and the city's marina district, the project included both design and construction of the four-lane, 2.3-mile roadway. The project scope included 16 bridges and construction of a 2,900 linear foot cut and cover tunnel with open depressed roadway sections on each end. The tunnel, which goes under Route 30 and a residential area, required the relocation of numerous utilities including water, steam, cooling water, sewer, gas, electric, and telephone. An existing sewer pumping station was demolished and relocated off the right-of-way of the highway. Environmental mitigation measures, a landscaped park, a pedestrian bridge, widening and resurfacing of several local streets, and the demolition of several city blocks of residential housing as well as portions of the Atlantic Electric power facility were also performed during this fast-track construction program. The program was successfully completed in 2001 (see Figure 4.14).

As part of the planning process, we (the design/builder) were responsible to develop a cost and resource-loaded computerized CPM payment schedule. This was unique, as the schedule was to be used as the basis for progress monitoring and for monthly progress payments. This necessitated planning for the integration of both budget and schedule control in one management system. A CPM payment schedule had never been developed before for a program of this complexity and size.

FIGURE 4.14
Construction of the Atlantic City/Brigantine Connector.

Several factors were considered during planning and development of the payment schedule including:

- Utilization of the schedule for payment. This was of particular importance because the execution phase of the program would generate up to $11 million per month in revenue and associated expenses. Proper scheduling and cost allocation would ensure that cash flow would be adequate to meet anticipated expenses.
- Utilization of the schedule for progress monitoring. Because of the urgent need for the roadway by the local gaming casinos, the program charter specified an aggressive timeframe: 13 months for design and a 28-month construction period. To "encourage" on-time completion, the owner set liquidated damages at $1.7 million for each week of delay.
- Utilization of the schedule as a program management tool: During initiation it was understood that such a complex and aggressive construction program would require the use of state-of-the-art management techniques such as what-if analysis, resource-leveling, and schedule

compression. The state-of-the-art tool, at the time, was PC (personal computer)-based CPM scheduling software. The systems had developed to a point where they were easy to use and provided quick and accurate results. Both were critical for this type of scenario planning.

Initially, we considered outsourcing the development of the payment schedule to a qualified minority contractor. Primarily because of the prohibitive cost to subcontract the service ($550,000), and the critical nature of the planning task, we decided to develop it in-house. We developed the payment schedule based on the concept plans provided by the owner and the detailed parametric cost estimate we prepared early in the planning process. During the development of the preliminary payment schedule, means and methods for many work items had not yet been established, so it was developed in a general format. Each of the 16 bridges, for example, was scheduled with only basic component descriptions: piles, foundations, substructures, structural steel or AASHTO [40] beams, superstructure, and appurtenances. The idea was that once detailed planning was complete and means and methods were established, it could be revised to reflect the added detail. The payment schedule was divided into four major work packages: planning and design (including design deliverables), permitting, construction procurement (including the submittal review cycle), and construction. The preliminary payment schedule contained 1,460 activities and 2,374 activity relationships.

The planning and design work package was scheduled with effort-driven activities using the New Jersey Department of Transportation Project Delivery Process Standard [41] for activity sequencing. Because the schedule was to be used for payment purposes, all of the design professionals' typical work tasks had to be included. This proved challenging for items such as RFI (request-for-information) reviews, and the creation of field drawings that were easy to resource and cost allocate but difficult to schedule because of their random nature. To resolve this problem, we scheduled these types of tasks with large float values.

The construction phase work packages were scheduled with task-driven sequence logic that had been determined during preparation of the parametric estimate. It was known by the team, early during the planning process, that finishing the program on time would depend on three work packages: relocation of utilities totaling $20 million in contract value and construction of the tunnel and boat sections, which included placing 125,000 cubic yards (cy) of concrete and over 650,000 cy of earthwork.

All activities were also resource allocated which included personnel, equipment, and materials. For the earthwork activities, a fourth resource category was added: quantity of earth-moving material. Although the program was complex in many ways, the program team viewed it as fundamentally a "big earthmoving job." Even before any formal scheduling had been done it was obvious that the critical path of the project would run through the earthwork activities.

There were six major earthwork activities: excavation of the tunnel (295,000 cy), installation of wick drains [42] (1,150,000 lf), placement of the roadway embankments (650,000 cy), placement of surcharge (40% of the embankment height), preloading of the surcharge for six months, removal of surcharge following a six-month preload period, and final grading. Scheduling of the earthwork tasks required the consideration of several critical factors. It was estimated, for example, that the most efficient production rate for each fleet of earthmoving equipment was 1,500 cy/day. Continuous operations were a must, as cost and schedule constraints meant that the equipment fleet could not be idle during the six-month embankment preload period. The correct number and size of the equipment fleet also had to be determined. The team worked in a collaborative way to find the best approach to addressing these considerations:

- For continuous equipment operation, the team did "what-if" type scenario planning on several different earthmoving sequences until the best solution was found. It was determined during this process that the wick drain operation was not on the critical path. This was a surprise given the amount of work involved with that operation.
- The schedule was resource-leveled based on the earth-moving production limit of the 1,500 cy/day. Resource-leveling assists in scheduling resources consistently throughout a program and ensures that they are not overallocated. Our CPM software system (Primavera P3) could automatically level resources based on resource limits. To level resources, the system would delay tasks with large float values or change task dependencies. Because these changes were done automatically by the software, it was critical that each leveling result was reviewed by the team. The speed, ease of use, and accuracy of the results proved indispensable, as they enabled us to investigate over 100 different scenarios.
- "What-if" analysis was performed on the schedule to determine the most efficient and cost-effective size of the equipment fleet. Variables

included the number of equipment fleet headings, individual equipment production rates, individual equipment rental rates, and the number and composition of each equipment fleet.

The final solution resulted in the division of the project site into five earthwork areas. Three equipment fleets would be utilized, one to excavate the tunnel and two to place the embankment and surcharges (preload and unload). It is shown in schematic format in Figure 4.15.

Using the earthmoving sequence logic as the driving factor, the team then planned the timing, and resource allocation, for the remaining construction tasks. The construction tasks were also all cost allocated based on 37 categories of lump sum payment items and a schedule of values totaling 1,289 items. In all, it is estimated that over 1,300 person-hours were spent planning for, and developing, the payment schedule.

Once accepted by the owner, the schedule was used as the basis for monthly progress updates. Because the updates served two purposes, providing a forecast of contract milestone dates and progress payments, establishing an accurate measurement for partially completed work was critical. The team had to determine which progressing technique should be used, either physical percentage complete, percentage of cost expended, or the actual number of days remaining. Physical percentage complete is a

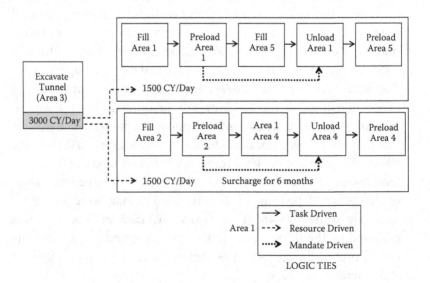

FIGURE 4.15
Earthmoving logic diagram.

progress metric determined by inspecting the work and using quantities to determine how much of a task is done. Percentage of cost expended takes advantage of the fact that each task had been cost allocated and assumes that cost and effort are in alignment. The third progress technique, the number of days remaining, uses actual or anticipated production rates to estimate the number of working days remaining to complete the task. We knew that what seemed like a relatively minor detail could have a great impact on the monitoring and control phase. To illustrate this impact, the invert slabs for the tunnel can be used as an example. Using Scenario 2, 30% of the slab's rebar installed for demonstration purposes, the progress calculated for the invert slab activity varies considerably depending on which technique is used to update its component tasks. It is calculated at 25% based on cost expended, 57% based on physical percentage complete, and 68% based on the number of days remaining. Progress payments for the month, depending on which progressing technique was used, could vary from $44,734 to $102,461 for just this one activity. Now that is quite a range!

We realized during the planning process that this variation could have a dramatic effect on the monthly payment amount and, more critically, on whether the updated schedule accurately forecast on-time completion of the contract milestones (see Table 4.5)

The program team chose to standardize on the days-remaining technique for progress monitoring as it was agreed that this method in general more accurately reflected schedule progress. This also tended to result in "overpayment," but in the collaborative environment that we were in it was agreed this was acceptable as it provided the contractor with positive cash-flow.

During planning, the team also set as a standard the mandate for a recovery schedule if the monthly updates showed the completion date falling behind by more than 2% of the remaining project duration. Recovery schedules were adjustments to the updated payment schedule, through either schedule logic revisions or duration acceleration, which eliminated any forecast delays to the completion date. A narrative explaining the adjustments to the payment schedule had to be provided with each recovery schedule outlining the steps to be taken to ensure the project would be completed on time. Explanations included items such as adding additional resources to accelerate activities on the critical path, working additional hours, working through holidays and weekends, and revising the overall sequence logic of the payment schedule to adjust the critical path.

TABLE 4.5

Progressing Techniques

		Qty/		Effort	Duration	
Task	Qty	Phr	$/Phr	Person-Hours	Days	Costs
Pour Mud Mat	67	0.6	91.31	112	2	$6,118
Erect Side Forms	1,260	6	9.78	202	7	$12,323
Erect Bulkhead	560	3	16.25	179	3	$9,100
Install Rebar	97,136	300	0.59	324	5	$57,310
Pour Structural Slab	934	5	98.45	173	1	$91,952
Strip Side Forms	1,260	29	2.11	43	2	$2,662
				1,033	20	$179,465

ACTIVITY: INVERT STRUCTURAL SLAB

PERCENT COMPLETE FOR ACTIVITY

Scenario 1 - 50% complete stripping side forms	97.9%	95.0%	99.3%
Scenario 2 - 30% complete installing rebar	57.1%	67.5%	24.9%
Scenario 3 - 15% complete erecting side forms	13.7%	15.3%	4.4%

PROGRESS PAYMENT FOR ACTIVITY

Scenario 1 - 50% complete stripping side forms	$175,686	$170,492	$178,134
Scenario 2 - 30% complete installing rebar	$102,461	$121,139	$44,734
Scenario 3 - 15% complete erecting side forms	$24,661	$27,368	$7,966

Both the payment and recovery schedule techniques proved to be important tools for monitoring and controlling the program's schedule and budget. Implementing these state-of-the-art planning processes required a great deal of effort from everyone on the program team. The end result was that a complex program, that experienced many bumps on the road, was completed on time and on budget. On July 27, 2001, the Atlantic City-Brigantine Connector had its grand opening celebration which included a tunnel walk and festivities that were open to the public. The tunnel currently carries approximately 25,000 [43] vehicles per day.

4.3 CHAPTER SUMMARY AND KEY IDEAS

4.3.1 Chapter Summary

For construction programs the majority of all major decisions should be made during the planning process. It is during planning where the key management baselines and the rules for engagement will be set for what, by nature, will be a raw and aggressive execution process. Deferring these vital decisions and critical choices until construction will result in seat of the pants type decisions and unfortunate results. Yet often, for various reasons, the correct amount of effort is not put into planning. And by failing to prepare properly, you are preparing to fail.

4.3.2 Key Ideas

1. The program management plan is a stand-alone document, which sets the monitoring and control structure for the strategic goals. In construction, it should contain component plans to create the master schedule; create the master budget; develop safety, quality, risk, change, and communications management plans; establish the rules of engagement for the contractors; and develop a transition plan. The program management plan should also include a program work breakdown structure (PWBS) which will detail the general scope established in the program charter.

2. In construction, creating the master schedule is the most critical of the planning processes. Both the design and construction phases should be planned utilizing the CPM scheduling technique with resource-leveling. Because of the "time is money" concept, setting the program time line is a prerequisite to establishing the budget.

3. Creating the master budget will rely on both parametric and analogous cost estimating for direct costs, categorizing and then forecasting incidental costs, and the setting of contingencies for unknowns. A collaborative effort between the design and construction professionals is an absolute necessity in order to create a comprehensive master budget that is acceptable to the owner.

4. The quality management plan describes the program's strategic approach to ensure the delivery of high-quality construction projects and ultimately its benefits to the stakeholders. The main goal of quality management planning is to avoid rework and failures. In construction even minor defects may require rework and in the worst case, failures. Because the cost of failure is so great, set procedures have been mandated in the construction industry for materials testing and inspection of workmanship.

5. Program risks are uncertain events or conditions that, if they occur, could have a positive or negative impact on an important objective. The risk management plan will include processes for identifying,

analyzing, and planning responses to risks. Both the probability and impact of each risk must be taken into consideration when developing responses. Responses must be cost effective, time sensitive, realistic, and have buy-in from the program team.

6. Safety planning involves three steps: establishing a programwide safety policy, determining safe work practices specific to each project site, and establishing an enforcement mechanism. The United States Occupational Safety and Health Administration (OSHA) mandates and enforces construction safety standards. It also provides safety training, outreach, education, and assistance, which all should be leveraged when developing the safety plan.

7. Effective communication, both formal and spoken, is a key to any successful program. On the program level communication is critical as it enhances collaboration. Effective teams are characterized by trust, respect, and collaboration. Internal communications should focus on collaboration and accountability, and external communications should focus on transparency.

8. Construction is different from other programs in that in most cases a major contributor, the construction contractor, is not at the table, either during the creation of the program charter or the creation of the program management plan. It is therefore critical during the planning process to establish rules of engagement to integrate the contractor's work with the rest of the program. Establishing the rules of engagement will depend to a great deal on the project delivery method(s) chosen, the management style of the owner, and mandates established by federal or state statutes.

9. During planning, it is critical for the program team to look forward to the transition of construction to operations. This is done by making sure the plans and specifications are implemented as intended and the completed projects are functioning correctly. It is during planning where considerations such as life-cycle costs, ease of operation and maintenance, training, requirements for warranties, and record documentation are vetted and solutions agreed upon by the program team. This will result in a smooth transition to operations, and if done correctly, continuation of the program benefits throughout its useful life.

5

Execution Process

5.1 INTRODUCTION

During program execution the projects are constructed and the program's outcome is created. It is through the proper implementation of the program management plan that the program's benefits will be realized. As elements of the program management plan are executed they may generate change requests. When change does occur, the monitor and control processes developed during the planning phase will be leveraged to identify the change and, if required, to take corrective action and remain on plan.

All construction programs are executed in three discrete phases: design, procurement, and construction. Each of the phases can be considered different projects within the program because each produces a specific outcome in a defined period of time. The timing and interrelationship between the phases will vary depending on the project delivery method. Proper implementation of the construction management (CM) execution process ensures that projects remain aligned with both tactical and strategic goals and, ultimately, the program produces its expected benefits and value.

5.2 CONSTRUCTION EXECUTION PHASES

5.2.1 Design

By nature construction projects and programs pose unique technical challenges. Each construction project must be designed to produce an output and contribute to the program's outcome and benefits. During the initiation and planning processes concepts are established to achieve these goals within the constraints of time, scope, budget, and quality.

It is important during the design execution stage that both schedule and budget are continuously monitored, and corrective action taken as required to stay on plan.

The best way to illustrate this is through example, so consider the LIPO program. The design of Metro was specifically targeted to reduce the amount of ammonia and phosphorus in the plant's effluent, which is directly discharged into Onondaga Lake. During initiation, it was determined that these two pollutants were a major cause of the lake's poor water quality. Nitrogen in the form of ammonia (NH_3) is a major killer of fish when present in high concentrations. Phosphorus poses an indirect threat to both the aesthetics of Onondaga Lake and to human health by delivering excessive nutrients that can promote the growth of algae. Algal blooms contribute to a wide range of water quality problems by affecting the potablility, odor, and color of the water. Other projects in the LIPO program focused on different technical issues related to Onondaga Lake's poor water quality. For example, several projects were designed to eliminate "floatables" and other projects to capture or store stormwater to reduce the number and amount of direct discharges of untreated waste into the watershed.

The program's technical goals had to be frequently monitored to ensure both the project outputs and the program outcomes could be achieved. For the projects, where the use of one-of-a-kind or state-of-the-art technology was often employed, this required precise technical specifications, control of construction means and methods, a rigorous start-up and testing process, and explicit performance requirements. For the program as a whole, this required concentrated monitoring and control measures for each construction project and a sophisticated, state-of-the-art system for the continual testing, and then computer modeling, of the quality of Onondaga Lake's water. Critically, both had to be done simultaneously, and the team had to be positioned to effect change if required.

In addition to solving the technical challenges of the projects and program, the team also had to monitor and control the budget and schedule as they passed through the gates of conceptual design, design development, and construction documents. At each design gate both the schedule, and then budget, were measured against their baselines. The schedule was tackled first as the confirmation of the program timeline was a prerequisite to determining the budget status. Both the design and the conceptual construction schedules were scrutinized at each gate. For the design execution process the task was to ensure that the intermediate milestones were achievable and that procurement and construction could start on time. During this process, design concepts, regardless of their technical merit, were rejected if they could not be implemented within schedule. Once the schedule was confirmed the budget was analyzed. Both parametric and analogous estimates were performed at each design gate and corrective action was taken to bring the program back within budget. The team used three separate techniques to make that happen: constructability analysis, value engineering, and scope reduction.

5.2.1.1 Constructability Analysis

The term *constructability* defines the ease and efficiency with which a project can be built by the contractor. The more "buildable" a project is, the less it will cost. Constructability is also a reflection of the quality of the design documents. If the design documents are difficult to understand and interpret, the project will be more difficult for the contractors to build. Constructability analysis is best performed by the construction professional, or by the trade contractors, as they will have the specific job knowledge and wherewithal to do it best. This will also allow for a review of the contract documents by a fresh set of eyes, which is especially helpful in housekeeping-type tasks such as correlating construction details with the technical specifications and plans. Also the construction professional or trade contractors are more likely to introduce innovative construction methods than the design professional as they are closer to current advances in their trade. If done correctly, constructability analysis has the potential to save both time and money. A good example is the construction of the concrete ceiling slab for the mechanical process piping gallery on the Metro project. The original design called for the ceiling to be cast-in-place concrete. Working closely with the design professional, the construction program manager proposed a design solution that was much easier to construct. The overhead concrete slab also significantly reduced the amount of time and effort required to assemble the mechanical process piping below. The idea was to use precast concrete slabs instead of cast-in-place concrete. This eliminated the effort involved with shoring and forming the soffit for the slab and the time required to cure the concrete (28 days per pour) before the shoring could be removed. It also allowed for the preassembly of the mechanical process piping which then could be picked up and lowered into the open gallery as illustrated in Figure 5.1.

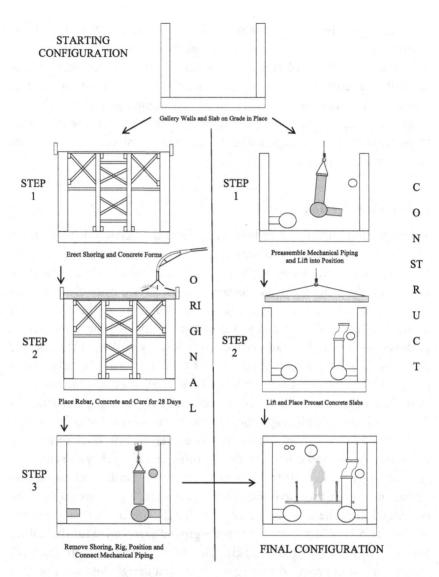

FIGURE 5.1
Constructability analysis example.

Because the precast slabs would become a permanent part of the structure, implementation of the concept required a significant amount of time and effort by the design professionals to incorporate it into the design. Taking that into consideration, and both cost and schedule savings, a cost–benefit analysis was performed and the idea was implemented. It was later determined that without this change in construction means and methods, the project (and program) could not have been completed on time (see Figure 5.2). More on that later in the case study at the end of this chapter.

Other examples of similar constructability suggestions that were implemented and saved significant time and cost from the original design concepts from other programs included:

- Increasing the thickness of a roadway tunnel's invert slab from 5 to 7 feet to reduce buoyancy and eliminate the use of tie-down anchors. This saved both time and money.
- Flattening the profile of a roadway tunnel to eliminate a pumping station. This saved both time and money.
- Using segmented concrete AASHTO beams instead of curved structural steel girders on a bridge superstructure. This saved both time and money.
- Using the cut and cover technique instead of soft ground tunneling on an underground wastewater storage facility. This saved an estimated $24 million.

FIGURE 5.2
Construction of the BAF gallery.

- Selectively using bolted connections instead of welded connections on the structural steel frame for a new building. This saved both time and money.

5.2.1.2 Value Engineering

Value engineering is similar to a constructability analysis in that both are efforts to find a less expensive way to do the same thing. In fact the terms often get used interchangeably which can lead to confusion. The difference is in the approach. Where a constructability analysis reduces costs by finding an easier way to achieve the same thing, value engineering reduces cost by finding design solutions that provide the necessary function for less cost.

The concept of value engineering evolved from the work of Lawrence Miles [44] who, in the 1940s, was a purchase engineer with the General Electric Company. This was during World War II, and there were shortages in steel, copper, bronze, nickel, bearings, electrical resistors, and many other materials and components. General Electric wanted to expand production of turbo superchargers for the B24 bomber from 50 to 1,000 per week. Miles was assigned the task of purchasing the materials to make that happen. Often he was unable to obtain the specific material or component specified by the designer so Miles would purchase a substitute. Where alternatives were found they were tested and approved by the designer. Miles observed that many of the substitutes were providing equal or better performance at a lower cost and this evolved into the concept of value engineering.

In construction, the value engineering process often takes on a negative connotation. In my opinion this is partly due to "pride of authorship" on behalf of the design professionals and is partly due to the often poorly conceived value engineering suggestions that are not equal in quality, or to the level of performance of the original design concepts. I once had an architect proclaim in a job meeting, as an example to help avoid such poor ideas, that "A wood stove is not a good value engineering idea for replacing the geothermal heat pumps." Despite its challenges and drawbacks, if done correctly, value engineering can result in significant cost and schedule savings without reducing quality. There are several things that need to be deliberated by the program team when properly studying a proposed value engineering suggestion:

- Does the suggestion satisfy the needs of the project or program?
- Is it a technically adequate solution?

- Is it in compliance with governmental regulatory requirements such as building codes and permits?
- Is it of equal quality?
- Will it have equivalent functional performance?

If these criteria are met then the next step in the process is to consider the cost of implementation. Implementation costs may include capital costs, life-cycle costs, redesign expenses, environmental impact, and other consequential costs such as schedule damage that are due to impacts or delays. Capital costs include the initial cost of construction, design, and procurement. Life-cycle costs include operation, maintenance, repair, salvage, disposition, decommissioning, sustainment, and replacement. It is critical that a comprehensive list of all implementation expenses is prepared so a proper cost–benefit analysis can be performed.

The value engineering process can be implemented both on component projects and at the program level. An example of the effective use of value engineering on the program level was the "greening" of the Onondaga Lake clean-up program. During wet weather events the stormwater from the city of Syracuse flows directly into the sewer system. These events overload the capacity of the Metro plant resulting in the direct discharge of combined sanitary flow and stormwater into the local tributary waterways of Harbor Brook and Onondaga Creek. These events are known as combined sewer overflows (CSOs), and they contribute significantly to the poor water quality of Onondaga Lake. The original solution to this problem called for the construction of large underground storage facilities to hold the CSOs until the wet weather event passed, and the Metro plant had the capacity to accept and treat the waste. These became known as "grey" infrastructure projects, and a total of $215 million in capital cost was spent on just three of the larger projects, which when combined, have the capacity to hold 212 million gallons of CSOs. More than halfway through the program, an idea to take a different approach arose. Instead of storing the CSOs for later treatment at Metro, the focus would be on eliminating the events in the first place. The approach was to use "green" infrastructure for capturing stormwater runoff in a sustainable and natural way. Over 100 projects were designed using green technologies, such as rain barrels, green roofs, rain gardens, porous pavement, bioswales, cisterns, and urban tree planting. These "green" infrastructure projects were estimated to be able to capture 109 million gallons of rainwater per year. A cost–benefit analysis was performed, and it was determined that the new approach would not add to the capital construction cost of the program and would result in $20 million in savings in operation and maintenance costs [45]. When implemented, the solution was so effective that the official name of the program was changed from LIPO to "Save the Rain."

On the project level the value engineering process is not generally applied to change the approach of the design but rather to find a less expensive way to implement the approach. A good example was the use of flexible plastic instead of rigid metal conduit for the electrical embeds in the concrete slabs at the Metro plant. At the time, now over 10 years ago, this was a novel idea and it was not well received by the design professionals. We argued that the function of the conduit was similar for plastic or metal as both would be embedded in the protective environment of the concrete. The advantages were significant, as we estimated a total cost savings of $65,335 (2004 dollars) based on ease of construction and material costs. After extensive deliberation, which eventually required the involvement of the owner, the idea was successfully implemented. It is interesting to note that, today, it is standard practice to use flexible plastic conduit embedded in concrete slabs.

5.2.1.3 Scope Reduction

Reduction in scope should be the last resort in an effort to reduce program cost. During the initiation process the general scope was established, and agreed to, by both the program team and its stakeholders. It is, therefore, imperative that careful consideration is taken when reducing a program's scope as a result of change. There are significant barriers against scope reduction that will limit the available options. The biggest barrier may come from the owner who based the decision to proceed on a general understanding of the program's scope. The need to reduce scope may be considered as a sign of poor planning and this may result in a lack of confidence in the team. An important stakeholder may also become unhappy if a scope item that is important to him or her is reduced or eliminated. There is also a limit to the amount of scope the team can reduce and still deliver on the program's benefits. This is known as the "whole baby effect" [46] where just like Solomon, the team will hear arguments that "You can't cut the baby in half and still have a valuable, or even functioning, program."

However, scope reduction may become an absolute necessity to save the program. Under such a scenario the best approach is first to prioritize each scope item by how critical it is to the delivery of the program benefits. This is best done through a brainstorming session where the diverse views of the program team can be vetted. The result of the brainstorming session will be an agreed-upon list of prioritized scope items that can be used later on in the decision-making process. The team should then reach out to critical external stakeholders to get their input and buy-in by contacting them and conducting interviews. The third step is to consider cost. When making these decisions it is important that accurate cost estimates are prepared for each scope item. The negative consequence of poor estimates may be that affordable scope items are left out or, conversely, items are left in that later become budget busters. Both hard and soft costs must be taken into account. Any significant reduction in scope should result in an incremental decrease in program soft costs. This does not happen automatically as the program team will need to look at staffing, and other associated soft costs, and make the sometimes difficult decisions to reduce costs there as well. The fourth step is to consider schedule impacts. A significant reduction in scope may result in the reduction in the program duration, and hence addition cost reduction, if the scope item is on the critical path. In these situations, the program team should be aware of the concept known as Parkinson's law that states, "Work expands so as to fill the time available for its completion"[47]. In other words, it is quite possible, without the team's proactive intervention that a reduction in scope will not result in all potential cost savings by reducing the schedule.

5.2.2 Procurement

During initiation a project delivery option is chosen that determines the program's final organizational structure. Project delivery options are methods by which "delivery" risks for the performance of the design and construction phases are transferred from the owner to another party or parties through the procurement process. The project delivery method will determine the structure for team accountability and set the rules of engagement for the team. As discussed in Chapter 2, project delivery options vary between the full separation of design and construction (design–bid–build) to complete integration of the phases (design–build). Regardless of the delivery option or type of work (public or private), there must be a separation of the responsibility for design from construction. This is done to protect the public's safety, health, and welfare. In the United States this requirement is regulated through each state's education laws and the licensing of the design professions. In New York State, for example, these laws require that all plans, drawings, and specifications relating to the construction or alteration of buildings or structures must be filed with a building code official and must be stamped with the seal of an architect or professional engineer (N.Y. Educ. Law § 7307 and Title 19 NYCRR Part 1203.3(a)(3)(1)).

An architect or professional engineer makes three distinct representations when stamping a document:

- First, a design professional's stamp confirms that the document was personally prepared by, or prepared under, the direct supervision of a specific individual and that that individual has accepted complete responsibility for the information contained in the stamped document.
- Second, a design professional's stamp affirms that the individual possesses the training, experience, and skills necessary to perform the scope of work encompassed in the document. By stamping a document, the design professional represents that the scope of work is within his or her "scope of competence."
- Finally, a design professional's stamp represents that the document conforms to the standards and requirements of the laws and regulations governing the practice of the applicable design profession.

For the "traditional" project delivery method of design–bid–build the implementation of these critical regulatory criteria is straightforward. The traditional method is purposefully structured to separate the responsibility for design from construction. This requires the procurement of independent vendors for the design and construction phases. Under this approach it is best to procure the design professional based on qualifications alone, and the contractor based on qualifications and price (best value). For the architect or engineer, because of the importance of his or her work for public safety and well-being, cost should not be considered when making the initial selection of the best or most appropriate provider of the professional services. This should be obvious, as no one would choose a surgeon based upon the doctor's willingness to perform an operation at the lowest cost. Negotiation of the cost for services should follow selection.

Application of this principle becomes more complicated with the use of alternative project delivery methods such as design–build or CM-at-risk where the services of the design professional are strategically aligned or even merged with those of the contractor. The intended advantage of these approaches is to provide a single point of accountability. However, it is important to note that this is for delivery risks only. According to law, the professional responsibility for design must still be separate from construction. It is often difficult in practice to distinguish between delivery risks and professional responsibility. Without the proper safeguards and oversight, this can lead to a mediocre design, or even unsafe practices, because the architect or engineer, if under direction of the contractor, may be more concerned with cost and constructability than the quality of design.

Regardless of the project delivery method chosen by the program team, the construction procurement phase will generally encompass the following eight stages:

- *Notices and Advertisements:* These are official notices of the owner's intent to search for qualified bidders for the work. The notice will include information about the program and will provide contact information for details about solicitation, rules for submission of bids, and awarding of the contract. For programs where public money is expended, additional information regarding governmental agency requirements, such as minority and women-owned business requirements, equal employment opportunity goals, prevailing wage requirements, and the use of indigenous labor, may also be included. It is important to note that there are specific legal requirements for these notices to ensure an open and fair bidding process.
- *Request for Proposals:* Commonly referred to as RFPs, these documents provide specific information regarding the professional or construction services desired. RFPs are written to ensure that the vendors respond factually to the identified requirements and should alert them that the selection process will be competitive. Well-prepared RFPs are done in a manner that allows for wide distribution and response. Construction procurement, either public or private, is expected to follow a structured evaluation and selection process that will demonstrate impartiality and fairness.
- *Pre-Bid Conference:* Once potential vendors do a preliminary review of an RFP they will invariably have questions regarding the procurement process, the construction plans and technical specifications, the general requirement provisions, and site conditions. The pre-bid conference is where these types of questions are answered or discussed with further clarifications determined as necessary. A properly

conducted pre-bid conference will also include an introduction of the program team including roles and responsibilities, an overview of the critical program constraints including budget and schedule, safety policies and procedures, quality expectations, and a tour of the construction site. The process will conclude with the preparation of meeting minutes and a listing of all attendees. The meeting minutes formalize answers provided; they can be used as a means to provide further clarification of issues raised during the meeting that could not be answered on the spot and will let the vendors know their competitors.

- *Pre-Bid RFI Management:* Pre-bid RFI (request-for-information) management is the process that allows vendors to raise questions and receive answers regarding the procurement process and contract documents prior to submitting a bid for the work. Pre-bid RFIs are also a method for vendors to clarify or resolve any perceived ambiguity in either the RFP or elements of the design. The pre-bid RFI management process is an important part of the design process as it provides the first external scrutiny of the plans and specifications. During the bidding process it is essential that the program team maintain a log of all RFIs and formally distribute both the questions and answers to all potential vendors (not just the vendor asking the question). This will demonstrate openness and fairness. The log should indicate the date the RFI was received, the subject, the identity of the requestor, the answer provided, and the date of the RFI response.

- *Addenda Management:* Addenda are written information adding to, clarifying, or modifying the bidding documents. Addenda may result from the RFI process, late changes in the program scope, or late corrections in the plans and specifications. Addenda become part of the contract documents when the construction contract is executed and must be planned, prepared, and managed with the same level of integrity as the original contract documents. This can be difficult to achieve during the often-expedited bidding process. Too many bid addenda, or a delay in the bid date to include addenda items, may indicate to potential vendors that the planning process was not managed properly. This, of course, is the absolute wrong message to send this early in the program and may result in elevated bid pricing.

- *Public Bid Opening:* The process for a bid opening will vary depending on the nature of the program, the project delivery method, and whether the work is public or private. In all cases there are common rules that should be followed to ensure openness and fairness. It is critical that these general procurement rules are established during the planning process and then implemented programwide for all projects. For construction, prospective contractors will generally submit sealed paper bids on the date and time established in the contract documents or as adjusted by addenda. Late bids should not be accepted under any circumstance as this can give an unfair advantage. All conformed bids are then given an initial cursory check by the project manager to ensure that all the required documentation is in order and then marked as such for a public reading that follows. To ensure openness and fairness, I am a proponent of public readings of all bid results for both public and private work, even though this is not generally mandated for private work. This will ensure the integrity of the bidding process and demonstrate the owner's commitment to the transparency. At this point in the process the bids are referred to as the "Apparent Bid Results." Immediately following the public reading, "Bid Tabulations," which show itemized bids, are prepared, printed, and distributed to those in attendance, along with the "Summary of Bid Openings" which shows the totals for the lowest and second lowest bidder for each project.

- *Recommendation of Award or Rejection of Bids:* Bids may deviate from the program estimates, sometimes significantly, for many reasons. The bidding environment, including the number and quality of bidders, macroeconomic conditions, and even the timing of the bid opening, can have a dramatic effect on the bid amounts. The owner has the right to reject all bids and if they come in significantly over budget this is the proper course of action. Under that scenario, the program team must go back to the planning process or in extreme cases, the initiation process, and start anew. I refer to this as the "control–alternate–delete" strategy. If the bids are in general alignment with the program budget, they then should be analyzed by the project manager for major deviations and irregularities and if none are found a recommendation for award is made to the lowest *responsive, responsible* bidder. A responsive bidder has fulfilled the requirements of the bidding process. A responsible bidder can perform the work as specified. Determining whether a bidder is responsible can at times be difficult, as it requires both objective (fact-based) and subjective (gut-feeling) judgment. This is critical as employing a contractor who is in over its head on any aspect of the work can not only have a negative impact on a project but can derail the entire program.

- *Contract Negotiations:* A meeting with the lowest responsive, responsible bidder will follow the recommendation to award and is sometimes referred to as a "post-bid conference." The post-bid conference is an opportunity to discuss the contractor's interpretation of the documents, means, and methods, and its general approach to the project. It is also the last opportunity for the program team to assess the contractor's willingness and ability to perform the work as specified. Contract negotiations follow and conclude with the acceptance and welcome of a new stakeholder and teammate.

5.2.3 Construction

During my training seminars on construction program management, I often joke at about three quarters through, that we are finally going to discuss construction! Notice that this is the case with this book as well. Many of the contractors at my seminars find that curious as construction is when their work is just beginning. This is by design and not by chance. The majority of critical decisions for a properly managed construction program happen during initiation and planning and not during execution. I would estimate the percentage of critical program decision points during the design phase alone at over 70%. This was recently corroborated in a comprehensive study done by the Pacific Northwest National Laboratory for the U.S. Department of Energy [48]. It concluded that the contractor is a principal decision maker for less than 15% of all issues related to energy efficiency in commercial buildings. This included the selection of the systems or means and methods for roofing, the building envelope, windows, space conditioning, water heating, lighting, appliances, and landscaping.

Construction is where the plans are implemented, monitored, and corrective action is taken as required. During construction change will inevitably occur, which may require adjustments to the plan. If the program has been governed properly from the start change will be limited and manageable.

The project delivery method that was chosen during the planning phase will have a significant impact on the program team's ability to manage change. The project delivery method sets the time of engagement and nature of the relationship with the contractor. It is critical to engage the contractor as soon as possible inasmuch as properly integrating their work with the rest of the team is a major component of successful construction execution.

5.2.3.1 Partnering and Project Labor Agreements

By the nature of the construction procurement process, the contractor is often late to the table, sometimes well after critical decisions have been debated, agreed upon, and put in place by the rest of the team. As a result, the contractor is frequently in the unenviable position of catch-up right from the start. It is, therefore, critical that the contractor quickly, but properly, integrate with the rest of the team. A formal partnering process is one proven way [49] to make that happen. Partnering is essentially a way to improve communications with the contractor. The main concept behind partnering is that trust, open communication, commitment, and a flexible attitude are all necessary to have a successful project. The partnering process is a way to attach a formal structure to that concept and provide a means to hold the team accountable to those goals.

The partnering process begins right after the contractor comes on board and is initiated when the program and project teams convene for a formal meeting, sometimes referred to as the "preconstruction workshop." The program team will often work with a facilitator to improve the quality and productivity of the meeting [50]. At the preconstruction workshop participants discuss and agree on the following:

- *Goals Statement:* This summarizes the goals established during the program charter but will now include input from the contractor.
- *Communications Plan:* This summarizes the procedures established in the program communications plan but will now also include details specific to the project and contractor.
- *Conflict Resolution Process:* This is a review of what is included in the special project provisions with discussion of specific details relevant to the project and contractor. It is important during this discussion that everyone agree on two fundamentals: that it is inevitable that issues will arise; and that all will endeavor to resolve the issues and conflicts to mutual satisfaction. Win–win solutions are not only tolerable, but preferred.

In follow-up workshops, the team addresses current concerns that have arisen on the project, using the forum to resolve issues, conflicts, and miscommunications. Project performance is also discussed relative to the established goals. Both the communications plan and conflict resolution processes may be fine-tuned in order to keep them effective.

Partnering has proven effective on my programs when implemented correctly, is accepted by all parties, and is championed by a strong advocate of the process. This is critical inasmuch as getting team buy-in to the concepts, especially mutual satisfaction in dispute resolution (win-win solutions), can prove difficult. This has become very apparent to me as I present the partnering concept at my training sessions across the United States. In New York City, for example, there is little appetite for partnering, especially for public projects where selection of the low bidder is mandated. The parties go into those types of projects preparing for a battle on the issues and assuming the worst of each other. I have heard that directly from both contractors and the design and construction professionals. Conversely, just 300 miles away in Syracuse, New York, on the same types of projects, partnering has become the norm, and in my opinion works exceedingly well.

Partnering is almost always considered an effective way to improve performance by those who choose to implement it on their projects and programs. Findings from a 1994 survey of 8,000 construction attorneys, design professionals, and contractors [51] confirm this:

- Design professionals indicated partnering was a "superior method" for achieving desired results.
- Contractors viewed partnering "as a highly effective vehicle for achieving a host of goals on construction projects."
- Design professionals and attorneys indicated a favorable to unfavorable experience ratio of five to one.
- More than 70% of all three groups predicted an increase in the use of partnering.

A flawed derivative of the successful partnering movement is the recent surge in project labor agreements (PLAs). Although in use since the early 1930s, PLAs are only recently becoming more the norm, especially on large or complex construction programs. A PLA is an agreement between labor and management that governs pay rates, benefits, work rules, and dispute resolution. These agreements are made between the owner and local trade representatives (not the contractors) and subsequently become part of the contractor's bidding requirements. According to those who support them,

PLAs help control costs and ensure that there are no disruptions to the construction schedule, or impacts on quality, from labor shortages, the use of untrained or unskilled workers, or strikes. Similar to the partnering process, a PLA includes procedures for problem solving and dispute resolution with a general agreement between management and labor to seek mutually satisfactory solutions to issues as they arise.

Because of the generally disorganized nature of the US construction workforce, a PLA through default is almost always an agreement between management and the local trade unions. Under the US National Labor Relations Act, construction contractors have the right to choose to unionize or not to unionize. The vast majority of contractors, more than 80% in fact, have voluntarily opted against unionization. But because, in my opinion, of the trade union's well-organized presence and local political clout, most PLAs require that contractors hire only union workers. Essentially, under those circumstances, no one represents the vast majority of construction workers or the nonunion contractors that employ them. And the groups that support the merits of nonunion contractors, such as the Associated General Contractors of America, do not have a seat at the table during the PLA negotiations, and this can lead to unfortunate results.

I have had significant experience with PLAs on my programs and it has not been positive. Although the partnering type aspects of PLAs can be effective, the process itself is marred with irregularities and shortfalls. Because most PLAs mandate the use of union labor they limit competition from nonunion contractors. And less competition generally means higher costs. Advocates for PLAs will argue that they actually save money by putting the program management team at the table with the trade unions where they can negotiate more favorable pay rates, benefits, and work rules than the individual contractors can. In fact on public programs it is often a requirement to show cost savings to justify implementation of a PLA. My experience is that these studies themselves are flawed as they often do not take into account the opportunity cost of the lost completion from excluding nonunion contractors and the significant additional risk associated with the program manager's responsibility for the dispute resolution process between the contractor and the labor unions. Essentially, in my opinion, PLAs require program managers to step in where they do not belong. Nonetheless, on the programs I have managed with PLAs, all of the required studies have shown significant potential savings by utilizing a PLA. I would argue that Norman Ralph Augustine [52] put it best when he wrote, "All too many consultants, when asked 'What is 2 and 2' respond 'What do you have in mind?'"

5.2.3.2 *Monitoring and Controlling Schedule, Cost, and Quality*

There are two primary objectives during the construction execution process. The first is completing each project on time and within budget while meeting established quality standards. The second is ensuring that the program benefits are realized and that it delivers on its value proposition. Achievement of these objectives must happen simultaneously, which takes a coordinated effort between the program and project managers. The PMI [6] summarizes the concept best: "Program managers ensure and check alignment; project managers keep the details of each project under control."

At the start of construction it is important to recall that during the initiation process the program team agreed to "plan to build," and then "build the plan." During the planning process the baselines were set for budget, schedule, and quality and the procedures to monitor and control them were established. During execution the team builds the plan. This means that major deviations from the plan are unacceptable. If change requires a significant deviation from the plan, it is the program manager's responsibility to determine whether the program is still feasible and justified. If it is not, the program should be terminated. The good news is that if the program has been governed properly from the start, most likely change will be limited and manageable. If fact, under that scenario, the execution process can be looked at as simply the management of change.

5.2.3.2.1 *Monitoring and Controlling the Master Schedule*

The master schedule developed by the program team during the planning process will contain the critical tasks and milestones for achieving the program's benefits, conceptual construction schedules for each project, and the interrelationships between the projects and other program components. The next step in the process is to update the conceptual construction schedules with the contractor's actual plan to move the work forward. The timing of this engagement will depend on the project delivery method and the sooner the better as the contractor's input to the construction schedule is a vital part of creating a realistic baseline. Because most construction programs comprise multiple projects with different start dates and durations, it is likely that the program team will be working with multiple contractors in developing the individual project schedules. Because of this it is critical that a standard process is developed for integrating the contractor's schedules with the master schedule. Recall from Chapter 4, this can be achieved with the development of special project provisions, included in the individual construction contracts, for planning and scheduling and the establishment of critical intermediate milestones and critical contract milestones. These rules of engagement will set the model for integrating the contractor's approach with the master schedule. However, the integration process still is often challenging as the contractor may have a different approach to the project than was anticipated when the conceptual construction schedule was developed by the program team. It is important during this early juncture that the integration process is managed correctly because it is the first important engagement with the contractor and it can set the tone for the rest of the project. From the perspective of the program team it should not matter what path the contractor takes to achieve the project schedule as long as it is proven feasible and reasonable (achievable), meets the major milestones, and does not have a negative impact on the critical intermediate milestones. The role of the program team should be to work with the contractor to make the contractor's plan work in the context of the master schedule. The attitude should be "trust with verification." Verification that the contractor's plan is achievable is the most critical part of the integration process. There are two major factors to consider when analyzing the contractor's schedule: the amount of float and the allocation of resources. To ensure this information is available, the special project provisions developed during the planning phase will include detailed instructions to the contractor on how to produce the schedule and prepare the expected deliverables. An example of an effective scheduling provision is included in Chapter 4.

The preliminary CPM schedule produced by the contractor will provide the program team with a treasure trove of information regarding activity float time that can be used to analyze the work plan. The amount of float, sometimes referred to as "slack time," is how long a task can be postponed before it affects other tasks or the project's finish date. There are two types of float time: free and total. *Free float* is the amount of time a task can be delayed before it delays a task that depends on its completion (i.e., a successor task). *Total float* is the amount of time a task can be affected until it directly delays the project finish date. Total float can be positive or negative. By definition a task with zero total float is considered a critical task as a delay in it will directly delay the project finish date. By examining the float time of individual tasks the program team can determine how reasonable and feasible it will be to complete the project on time under the contractor's work plan. For example, if the schedule contains any activities with negative float then it is by definition infeasible. If the schedule contains a large portion (i.e., over 40%) of tasks with total and free float values at, or near, zero, then that would indicate the schedule is unreasonable. The proposed schedule would leave little, or no, time to react to unforeseen conditions or other change events. The acceptable amount of float time in a schedule is subjective and can only be properly determined by an experienced and knowledgeable construction professional. Judgment is required based on the type of project, the project and program schedule constraints, the availability of resources, and the contractor's willingness and ability to perform.

The second way to determine if the contractor's work plan is achievable is to examine the allocation of resources over time. By having the contractor discretely load each task by its required resources, including personnel, materials, and equipment, the scheduling software can create resource histograms of the proposed expenditure of resources over the project duration. A good example of this technique can be taken from the Atlantic City/Brigantine Connector program for the scheduling of the Manitowoc 4100W cranes. These large 230-ton crawler cranes were a critical resource for us. At the time there were a limited amount of them available throughout the United States, making them very costly to purchase or rent. In addition, a specialized crew, and a second crane, was required to erect and dismantle them making mobilization expensive and time consuming. Furthermore, it was expensive to operate them, requiring a composite crew of an operator and oiler. We had requirements for them on several aspects of work including the placement of rebar and concrete for the tunnel, erection of structural steel, and AASHTO beams, the pile-driving operation for the 16 bridges, and the installation of the million linear feet of wick drains.

In order to ensure we did not overallocate the crawler cranes, we resource-loaded each task that required the Manitowoc 4100Ws in the preliminary schedule and the software created a resource histogram as shown in Figure 5.3. It revealed, to our surprise, that five of the Manitowocs would be required to complete the project as we originally scheduled. Not only was this one more than was available, the resource histogram also revealed the inefficient use of the cranes with much heavier usage during the summer and fall months.

This demonstrated to us that the original schedule, although it may have been feasible, was not reasonable. It also helped us see the forest for the trees, as we probably would have missed the overallocation and inefficient use of the crawler cranes. At that time we were more focused on the individual task schedules and not on the allocation of resources. Ultimately

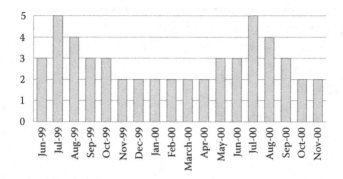

FIGURE 5.3
Resource histogram (peak usage) example.

we resource-leveled the schedule to limit the peak usage of crawler cranes from five to four.

This technique can be used in a similar fashion for any resource that is required on the project or program. Some examples are the following:

- The peak requirement for carpenters on a program. We used this information to help negotiate a PLA agreement.
- The total amount of concrete required for a project. We used that information to ensure availability and negotiate a better price.
- The amount of structural steel delivered to a project in a given week. We used that information to coordinate the deliveries with local traffic officials.

The process to integrate the contractor's work plan with the master schedule is interactive with analysis using the techniques above, and then using what-if type scenario planning as the schedule is revised. Depending on the sophistication of the contractor and its willingness to work with the program team, and the complexity of the program, the process can take several weeks or even months to complete. Once the contractor's schedule is successfully integrated with the master schedule, I like to have a "signing ceremony" to celebrate the important milestone and the hard work and collaborative effort that it took. At the signing ceremony, everyone is required to sign the schedule, including the construction professionals, the design professionals, and the contractor. This sends a powerful message early on that the team has worked together to come up with the work plan and that everyone has bought into it. The signed schedule is then displayed prominently in the construction

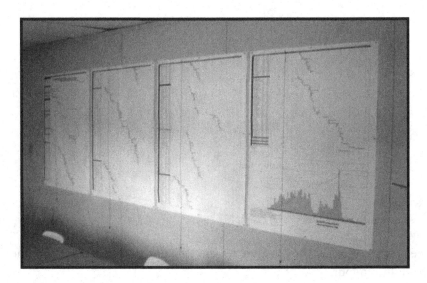

FIGURE 5.4
Integrated master schedule: Cayuga-Onondaga BOCES.

trailer (see Figure 5.4). This is a great motivator, as at each biweekly progress and coordination meeting, the team must walk by their signatures posted on the schedule.

Once the contractor's schedule is approved and integrated with the master schedule, it is used as the basis to monitor the progress of the work. In order to monitor the schedule properly both a short- and long-term view are needed. For the short-term view, the contractor will produce weekly "three-week lookahead" work plans. These work plans will be detailed, including tasks drilled down to daily activities, which are then integrated with the master schedule through the consistent use of activity descriptions and coding. The three-week lookahead schedules are updated each week by the contractor and submitted to the program manager for review. At the biweekly progress and coordination meetings schedule progress is discussed using these updated short-term work plans as the basis.

Once a month, the contractor will also be required to update the master schedule and submit it to the program manager for review. This provides the long-term view. The scheduling software will automatically forecast the intermediate and finish milestones based on the progressing information entered by the contractor. But be aware, as the old adage warns us, "Garbage in, garbage out." For the software to perform a proper analysis of schedule progress, it is critical that the progressing information is accurate and timely. The best approach is for the contractor and program manager to agree in advance on the activity progressing technique (percentage complete or days remaining) and collaborate on gathering and agreeing on the monthly progress data.

Because the schedule software was cost-loaded (or from the contractor's perspective, revenue-loaded) discretely by task, it can produce data on earned value as well. Earned value, simply stated, is what the contractor has achieved with the revenue it has been given. And because current performance is the best indicator of future performance, it is possible to monitor the schedule using earned value trend data. There are three critical variables in this type of analysis:

- *Budgeted Cost of Work Scheduled (BCWS, also known as Planned Value.* The planned cost of the work scheduled to be performed. You can see an example of this type of data displayed in the cost histogram on the bottom right of Figure 5.4.
- *Actual Cost of Work Performed (ACWP), also known as Actual Cost: the* cost of the work done to date. This is equivalent to the to-date amount of the contractor's current pay requisition.
- *Budgeted Cost of Work Performed (BCWP), also known as Earned Value.* The planned cost to complete the work that has been done.

With these data the scheduling software can calculate five performance metrics:

- *Schedule Variance (BCWP-BCWS):* This is a comparison of the amount of work performed compared to what should have been completed. A negative variance is a sign that the project is behind schedule.
- *Cost Variance (BCWP-ACWP):* This is a comparison of what the contractor should have been paid if it was on schedule to what it was paid. As most contactors like to get paid, a negative cost variance is a good indicator that the project is behind schedule.
- *Schedule Performance Index (SPI = BCWP/BCWS):* If the SPI is less than one then the project is behind schedule.
- *Cost Performance Index (CPI = BCWP/ACWP):* If the CPI is less than one then the project is over budget.
- *Cost Schedule Index (CSI = CPI × SPI):* The farther the CSI is from one, the less likely recovery will be effective.

These schedule performance metrics can be used in conjunction with the analysis of float time and resource usage to get a complete view of the status of the project or program. Monitoring the contractor's plan using these techniques is an effective way to measure contract performance. This has been proven on many of my projects and programs. However, the CPM schedule is simply a computer model of a much more complex activity. More subjective analysis, which requires judgment and past experience, must be used in conjunction with the software tools. This is especially true when change occurs that requires the team to deviate from the planned course.

It is critical for the program team to be proactive when a schedule impact is revealed during the monitoring process. The team's ability to control the consequence of an impact decreases with time. A project does not stop for impacts, and as it progresses there is less time remaining to adjust or recover. Bad news just doesn't age well. Once a negative impact to the critical path happens, there are just two scenarios: the project will either be delayed, or it will need to be recovered. Thus a negative impact on the schedule will almost always negatively affect the budget. Determining the best course of action is one of the most important decisions the program team will make during the execution process. There are many things to consider including the cost versus benefit of delay versus recovery, the impact of the project delay on other program projects or components, and the required level and availability of resources. We discuss these in more detail in the case study at the end of this chapter.

5.2.3.2.2 Monitoring and Controlling the Master Budget

As described in Chapter 4, the master budget is developed as a comprehensive register of all program budgets categorized as indirect costs, direct expenses, and contingencies. The master budget is used as a baseline to monitor cost and, if required, to initiate corrective action. Depending on the origination of the expense, timing, and scope, each budget category requires a unique cost monitoring and control system.

The indirect budget is used to support the program management office (PgMO). For these types of expenses the focus of the monitoring effort is on the rate of expenditure over time. Indirect costs start during initiation and end at the transition of the program to operations. Indirect expenses will span several projects and for some construction programs they may run up to a decade into the future. For programs supported by public money, the rate for reimbursement of indirect costs often has a fixed maximum threshold, usually as a percentage of direct costs. The combination of these two circumstances can make it challenging to manage indirect costs properly. The best approach is to focus on the largest category of controllable indirect costs. For construction programs this will, by far, be the cost of the management and technical staff including professional consultants and advisors. The key to success is to assemble an efficient program team and complete the program on time. An efficient program team will have the best people in the right positions. It is important to have the proper level of staffing at all times. Simply cutting staff to manage indirect costs is not a good strategy. The program-level contingency should be used instead. Reducing staff can have the unintended effect of increasing direct costs by lowering the level of oversight and the team's ability to take corrective action when needed.

The direct budget is used to execute the program. For these types of expenses the focus of the monitoring effort should be on change control. And because of the concept of "time is money" it is critical that a monitoring and control system be in place to react to change immediately. A relatively small change can result in a large cost overrun if it is allowed to affect the schedule's critical path and delay a component, a project, or the program as a whole. This is especially true in construction where time is always of the essence. The best strategy to be able to react immediately to change is to put in place an effective monitoring system and decision-making process. For the construction projects the responsibility for monitoring the work will fall on the design and construction professionals and may include a third-party inspection team. The most important component of the monitoring system is to have an experienced full-time team dedicated to each project that can recognize early on when a changed condition happens and be able to bring it to management's attention immediately. The change-order process established during planning and included in the construction contracts will then be used. There will be several key decision points when the change-order process is implemented. To ensure the process runs efficiently, it is important to establish decision-making tolerances based on level of responsibility. For example, an excessive cost overrun may be outside the range of tolerance for the project manager to resolve and require corrective action by the program manager. This hierarchical decision-making process should be well defined with "flags" and "triggers" to ensure quick reaction to any level of change.

For programwide issues the responsibility for cost monitoring and control will fall on the program management team. These types of issues also are associated with change events, but from a strategic rather than tactical point of view. These include items such as:

- Large project change events that have the potential to affect the program budget
- Interproject change events that have the potential to delay the program and affect the budget
- Major changes in the composition of the program or project management teams
- Change in the cost of resources
- Change in the influence and interest of major stakeholders
- Change in economic conditions
- Change in financial conditions

The risk management plan developed during the planning process will include strategies for dealing with these types of change events. The program contingency is used to offset the cost impacts. If the contingency becomes depleted, a request for an increase in the program budget is made. That decision is then made by the governing body based on the current feasibility and justification for the program.

5.2.3.2.3 Monitoring and Controlling Quality and Value

Projects produce outputs and programs produce outcomes. In order to realize the full benefits of a program the component projects must be built with quality and also contribute to the value of the program. Most construction programs deliver benefits incrementally through completion of the individual projects. The project teams are responsible to ensure that each project is built with quality and that they achieve the desired level of performance. The quality management plan will contain process and procedures for both quality assurance and quality control. The quality assurance process makes sure that the project team is doing the right things, the right way. The goal of the quality assurance process is to ensure that the standards, processes, and policies that were established in the quality management plan are in place and executed. The quality control process makes sure the results are what were expected. On most construction projects a third-party testing and inspection agency is responsible for the monitoring aspects of quality control and the project team is responsible to take corrective action as required. The technical plans and specifications will contain the quality standards, whereas the general conditions will outline the process for correcting defective work.

The program team is responsible for managing the group of projects in a coordinated and consistent way to ensure the delivery of the program benefits. During the execution process, the program team brings value by:

- Ensuring that shared resources are coordinated across projects
- Reacting to opportunities and risks
- Measuring and communicating the program status to stakeholders
- Reviewing major change orders and authorizing extra work
- Taking corrective action when projects do not meet their expected results
- Identifying environmental change that might affect the execution of a project or the delivery of the program benefits
- Ensuring that the program continues to be justified and feasible

During the execution phase the project teams focus on construction quality whereas the program team focuses on value and benefits delivery. The PMI [6] summarizes it best, "Program managers ensure and check alignment; project managers keep the detail of each project under control."

5.3 CASE STUDY OF AN EFFECTIVE MONITORING AND CONTROL PROCESS

When Benjamin Franklin coined the phrase "time is money" over 200 years ago in his *Advice to a Young Tradesman*, he introduced to the world the concept that time lost is money spent. As we have seen throughout the book nowhere is this concept truer than on modern-day construction programs. Time is money on construction programs because of the close interrelationship between schedule and cost. The following case study demonstrates how the monitoring and control process, and the schedule recovery technique, mitigated schedule impacts in a cost-effective way. I use the Metro project and LIPO program again as our example. In order to appreciate fully the scale and complexity of Metro, and to "set the stage" for the recovery efforts, a detailed description of the challenge follows.

The scope of the work and timeline for the start and completion of Metro stem from a 1998 amended consent judgment (ACJ) settling litigation between the State of New York, the Atlantic States Legal Foundation, and Onondaga County, in connection with alleged violations of state and federal water pollution control laws. The ACJ established criteria for discharge of effluent into Onondaga Lake including a limit of two mg/day for ammonia by May 1, 2004, and a limit of.12 mg/day for phosphorus by April 1, 2006. The ACJ effluent limits are some of the most stringent requirements in the United States.

To achieve these criteria and the aggressive timeline for completion, the Metro upgrades were designed as three separate facilities combined into one large complex. A biological aerated filter facility (BAF) to remove ammonia; a high-rate flocculated settling facility (HRFS) to remove phosphorus; and an ultraviolet disinfection facility (UV) to disinfect the effluent prior to discharge into nearby Onondaga Lake. Flow from Metro is conveyed to the complex through a new 130-million-gallon-per-day secondary effluent pumping station (SEPS). Onondaga County controls the new facilities and all of the systems of the existing 50-acre, 240-million-gallon-per-day plant from a new three-story plant operations center that was constructed as part of the project. Contributing to the complexity of the Metro project was its location on a brownfield site. Contaminated soils and groundwater deposited from a manufactured gas plant (MGP) operation had to be remediated. As a condition of a separate New York State Department of Environmental Conservation (NYSDEC) consent order, Niagara Mohawk Power Corporation (NiMo), the site's previous owner, was responsible for remediation of the 3.2-acre site. Through a negotiated agreement with NiMo for acquisition of the land, the county was responsible for the clean-up of the Metro site with partial reimbursement of the costs from NiMo.

Metro is the biggest complex of its kind in North America (see Figure 5.5) and the $132 million project generated a significant amount of construction work:

- More than 150,000 tons of MGP-contaminated soils were removed and 270 million gallons of contaminated groundwater were treated.
  - A total of 1,108 H 14 × 102 steel piles were each driven over 250 feet (over 50 miles in total).
  - A total of 28,500 cubic yards of structural concrete were placed.
  - $13.8 million of owner-supplied process equipment was prepurchased and had to be coordinated for fabrication, delivery, and installation with the installing contractor's schedule.

FIGURE 5.5
Plant operations center at Metro.

- There were over 466 trillion polystyrene beads installed, which contained explosive chemical residue, in the 18 BAF cells.
- Over 130,000 stainless steel nozzles were installed in the BAF cells to permit the treated effluent to leave the BAF cells while trapping the polystyrene beads for continued use in the treatment process.
- The HRFS required 100 tons of microsand, which is recycled continuously to the system.
- More than 790 linear feet of pile-supported, 72-inch prestressed concrete cylinder pipe (PCCP) force main and over 360 linear feet of pile supported 84-inch PCCP effluent pipe were installed.
- And more than 5,000 linear feet of additional mechanical piping were installed within the existing plant. The existing plant's boilers were replaced. The existing plant's electrical distribution and supply system were expanded.

The ACJ required the successful operation of a BAF pilot ammonia removal demonstration project by November 1, 1999, followed by submittal of approvable engineering reports and plans for the BAF to the NYSDEC by December 1, 2000. Although the ACJ milestone for submittal of approvable engineering reports and plans for the HRFS was not until

June 1, 2005, over five years later, the program team elected to pilot the HRFS simultaneously with the BAF pilot program. The pilot programs for both the BAF and HRFS proved effective and successfully demonstrated the new technologies. Based on the positive results, the program team decided to combine the HRFS and BAF into one facility which would give the county the option to achieve the ACJ requirements for both phosphorus and ammonia removal at the same time. It was also decided that a new UV facility would be included to replace the plant's existing, and dated, liquid sodium hypochlorite disinfection system. Combining the three facilities resulted in a significant reduction in project costs through the economy-of-scale principle and a significant reduction of the timeline for the entire program. It has been estimated that this choice, combined with the use of the new process technologies, saved over $70 million.

Once planning was complete the team moved directly into the execution phase. The design professionals knew they had to prepare immediately. Combining the facilities added significant complexity to the design process while at the same time reduced the timeframe. Despite the challenge, through the use of additional resources, the engineering reports and plans for Metro were completed on schedule. The notice to proceed on the first construction contract for the test pile program was issued in June 2001. Although still on plan, the remaining schedule was aggressive, allowing just 29 months for completion of the remaining bid documents; remediation activities; and construction and performance testing of the BAF, SEPS, and UV facilities; and an additional seven months for completion of construction and performance testing of the new plant operations center and HRFS facility.

To achieve the aggressive construction schedule, the project was fast-tracked, meaning that when site remediation and other site-work activities were underway, the bidding documents for the construction of the BAF, SEPS, HRFS, and UV facilities were not yet complete. In fact, during this timeframe, several changes to the final design of Metro were made, including the addition of an 84-inch, PCCP, pile-supported, underground bypass system. The bypass allowed the BAF to operate independently of the HRFS. Although the bypass added significant upfront project cost and negatively affected the schedule, it also added schedule flexibility by allowing the more critical BAF to go online first, prior to completion of the HRFS. Ironically hydraulic considerations required the bypass pipe to have the deepest excavation on site, which significantly delayed the remediation, site-work, and pile-driving operations. During the planning process the program team considered several different approaches for the execution of the construction contracts. The final solution divided Metro into 11 separate work packages. The basic sequence logic for the final solution is shown in Figure 5.6.

The contracts were structured so that the site-work contract, including the remediation of 154,700 tons of contaminated soil, was a prerequisite to the start of pile-driving and all successive facility work. This structure accommodated the special requirements for handling the contaminated materials, including health and safety regulations. This method limited risk by confining the remediation activities to one contract.

The remainder of the project was sequenced to coincide with the ACJ major milestones for the BAF and HRFS facilities. The ACJ included major milestones for both completion of construction and successful operation of both the BAF and HRFS facilities. Successful operation of the BAF and HRFS was defined by achievement of the set limits for ammonia and phosphorus in the effluent. The BAF facility was required to meet the effluent limits much sooner than the HRFS, thus the BAF facility became the critical path of the project (see Figure 5.6).

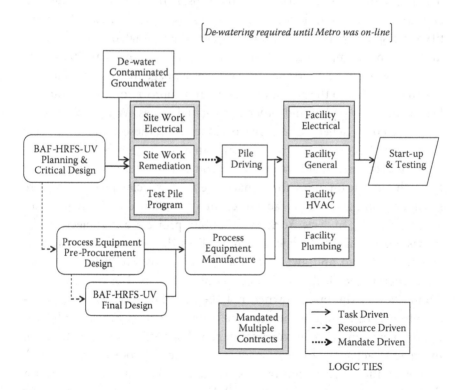

FIGURE 5.6
Metro basic sequence logic diagram.

The site remediation contractor began work in July 2001. The program team knew that the aggressive nature of the schedule left little float for delay in the start or execution of the work. Because of the direct relationship between the remediation and pile-driving operations, any slippage in the time to remediate the site could directly affect Metro's mandated completion dates. When developing the program management plan the remediation activities were identified as a high-priority risk because of the level of uncertainty in defining the actual scope for this type of work. Unfortunately, these concerns were realized when a number of unforeseen site conditions severely affected the progression of the remediation work. Excavation uncovered several large, below-grade, pile-supported concrete structures that obstructed installation of the supports for the excavation system. The concrete obstructions and the timber piles beneath them had to be removed prior to continuation of the work. Once excavation was under way, the poor consistency of the contaminated material made handling and disposal difficult, requiring the contractor to "amend" the material with imported sand prior to disposal. The sand and the overexcavation for the removal of the concrete obstructions increased the quantity for disposal at the landfill by 95% over the estimated amount in the unit price bid item, and the quantity of backfill by more than 200%. The situation was further worsened by a 60-day slippage in the start of activities that was due to the late development and approval of the contractor's health and safety plan and other contractual issues. Winter conditions also affected the schedule, and by November 2001, these cumulative delay events had already affected the critical path by eight months.

Because the major milestones were court mandated and enforceable through significant fines for nonachievement, special provisions were included in the prime contracts to ensure compliance with the project's schedule goals. In New York State, public projects are subject to the provisions of the Wicks law, which requires that separate contracts be competitively bid and awarded to a minimum of four prime contractors: general, electrical, HVAC, and plumbing trades. The Wicks law also stipulates that the public owner, not the general contractor, is responsible for coordination of the multiple prime contracts. Because of these requirements, many public owners in New York State use a construction manager to act as the owner's agent to schedule and coordinate the work of the multiple prime contractors. At Metro we were responsible for developing an integrated master CPM schedule for the program. The scheduling provisions focused on ensuring that the prime contractors provided detailed schedule information for their individual work activities. Each prime contractor was required to submit a comprehensive list of work activities, which included submission and approval of all project deliverables; all required tasks for the procurement of equipment and materials; all construction work tasks; and project closeout tasks, including punch list and testing activities. The prime contractors were also required to furnish a brief description of each activity, provide the activity duration, establish predecessor or successor activity relationships, and discretely load each activity for revenue, equipment, and manpower requirements. Based on the information provided, the construction manager prepared the integrated master CPM schedule that was used as the baseline to monitor schedule performance.

The types of contract requirements outlined above for CPM scheduling are common on large/complex construction programs such as Metro in the United States (see Figure 5.7). What was unique on the Metro program was the effective use of special provisions regarding recovery schedules. This provision read, in part:

> If in the view of the Construction Manager, the Contractor is in jeopardy of not completing the Work on time, or not meeting any

FIGURE 5.7
Metro exterior (foreground).

schedule project Milestone, the Construction Manager may request the Contractor to submit a recovery schedule. The recovery schedule shall show, in such detail as is acceptable to the Construction Manager, the Contractor's plan to meet all schedule project milestones, and that the Work will be completed within the time frame stipulated in the Contract Documents.

A recovery plan was further defined as an adjustment to the schedule logic or the acceleration of activities on the critical path, which would eliminate forecast delays to the major milestones. If a recovery schedule was requested by the construction manager, the affected prime contractor had to provide a narrative explaining the adjustments to its work plan that would be implemented to guarantee the project would be completed on time. Explanations could include items such as adding resources to accelerate activities on the critical path, working additional hours, working through holidays and weekends, change in means and methods, or revision of the overall sequence logic of the CPM to adjust the critical path. The specifications also stipulated that all payment to the contractor would be withheld if an acceptable recovery schedule was not provided within 30 days of the request.

The provision was careful to identify the construction manager, and not the prime contractors, as having the authority to request and implement a recovery schedule. This was done because of issues regarding cost versus benefit of recovery, contractual requirements such as the effect on liquidated damages, the potential impact of recovery efforts on the other prime contractors, and the potential impact of recovery efforts on existing plant operations. This special provision intentionally did not address the issue of reimbursement for the cost of implementing the recovery plan. Other contract provisions in the general specifications, including "changes" and "time provisions," addressed this issue. This approach was taken as it alleviates most of the negative conflict and finger-pointing that can occur in these situations and focuses the contractor instead on developing an effective recovery plan. The inevitable issue of responsible party, and cost of the effort, is handled as a separate issue later through the contract change order process.

As discussed above, the first major impact on the schedule occurred during the site remediation activities. Because of the uncertainty associated with this type of work, the construction manager decided to implement only minor recovery efforts during this phase of the project. The most significant recovery effort involved reimbursing the contractor to send

FIGURE 5.8
Excavation for Metro.

the contaminated materials to the landfill without amending the soils with sand. Although this resulted in a surcharge of $27 per ton from the landfill, the excavation production rate on the site nearly doubled, greatly reducing the impact of this delay (see Figure 5.8). Not only did this save time, but the cost of the surcharge was more than offset by the reduced weight of material entering the landfill (because no sand weight was added).

Once this recovery was implemented, the focus for the construction manager became the acceleration of successive work, particularly the 50 miles of piles that had to be driven. It was important to accelerate this work because work on the process facilities could not begin without the piles in place. After consideration of cost and risk factors, a recovery schedule was implemented as follows:

- The start of pile-driving operations was allowed to overlap with the finish of the site remediation activities, meaning that pile-driving activities began before the remediation work was complete. As a result, the pile-driving contractor was issued a change order for $149,281 to implement a health and safety plan to work on the contaminated site.

- The prime contractor agreed to increase the number of production pile-driving rigs from three to five at no additional cost. This required that the owner relax rules that stipulated the working distance between the pile-driving rigs and other activities. The contract required the pile-driver to maintain a minimum distance of 100 feet between the production rigs and other work activities, but this requirement was relaxed to 50 feet to allow for the additional cranes and the overlap of work activities.
- The prime contractor agreed to alter the specified means and methods for pile-driving at no cost. Although the specifications called for the use of a vibratory hammer, this was relaxed to allow a more efficient impact hammer method, which greatly increased production.

Work on the pile-driving contract commenced in February 2002, and was completed within the original contract duration of 150 days. The recovery efforts ensured that the original contract duration was maintained even though there were several modifications that directly affected the work:

- Unforeseen buried concrete structures
- Issues regarding coordination of work with Contract 2
- Impact and stoppage of work due to health and safety precautions that were related to contaminated materials handling
- Several changes in the design of the facility that increased the quantity of piles by 25,372 linear feet from the estimated amount in the unit price bid item

Had recovery efforts not been implemented, these impacts would have resulted in at least a 60-day extension of the 150-day contract duration.

Notice to proceed for the facility construction contracts was issued in January 2002, and the general contractor immediately began preparation of a CPM schedule. In April 2002, based on the analysis from the CPM schedule, which included the impacts from the site work and pile-driving delays, the general contractor submitted a request for a 13-month time extension. This was quite a setback for the project team. Even with the team's effort to mitigate the previous impacts on the schedule, the project still faced what appeared to be an insurmountable task. Was it even possible to recover from a 13-month impact on a project with an overall duration of 29 months?

Could the construction manager count on the prime contractors, owner, and design professionals to buy into any proposed recovery plan? How much would recovery cost and who would pay for it? Could it be done and still maintain the quality standards established by the project team? We took a very formal approach to answering these questions and better communicating the issues to the project team. First, we carefully performed a "what if" type analysis of the schedule to determine if recovery were feasible. If a recovery were feasible, then a cost–benefit analysis would be performed to determine if it was the best course of action.

Table 5.1 illustrates the issues addressed during this process, outlining the issues we addressed to prove a recovery was feasible, and Table 5.2 outlines the things that we considered when we performed the cost–benefit analysis to determine if it was justified.

For over four months the program team debated what course to take, given the identified issues and the cost–benefit considerations. During this debate, the most difficult concept for us to convey to the rest of the team was that regardless of which choice was made there would be a significant impact on the project budget. Time is money. Schedule impacts, delays, and the recovery of lost time all cost money. To better covey this message we performed detailed estimates for both the cost of a 13-month delay and the cost to recover the lost time. We presented the information to the team in graphical format as shown in Figures 5.9 and 5.10. The general specifications contained a "no-cost-for-delay" provision that some members of the program team felt eliminated most of the potential cost for delay. This debate focused on whether the no-cost-for-delay clause was enforceable. If it was it would preclude the prime contractors from recovering damages as a result of the schedule impacts from the delay in the site work activities.

Although much has been written about the enforceability of no-cost-for-delay provisions in the United States, it was still a dividing issue and definitely influenced the decision-making process. We argued that even if these costs were factored out, other savings, such as the elimination of extended project soft costs for the program team, extended time-related unit price items, and the ACJ penalties, clearly favored recovery.

Some project team members were also not convinced that the proposed recovery schedule could be implemented without having a negative impact on the quality of the work. We agreed with this assessment only if additional focus and procedures to ensure quality were not implemented during the recovery efforts. We argued that ensuring a high level of quality during recovery required that the design professionals, construction

TABLE 5.1

Recovery Feasibility Analysis

Issue	Analysis
Could we demonstrate to the owner that recovery is needed?	The best tool for this is state-of-the-art CPM scheduling techniques with real-time controls for progress monitoring. At Metro we used Primavera CPM Scheduling Software integrated with the Primavera Expedition Document Control System. We developed a detailed integrated baseline master CPM schedule that was updated with real-time (weekly) data provided by each prime contractor. Each week, based on this information, we provided a forecast of the ACJ major milestones. These forecasts highlighted the specific areas where recovery was needed.
Could each of the prime contractors demonstrate and commit to a recovery schedule that they believe is achievable?	The strategy at Metro was to work actively with the general contractor to establish a recovery plan that was feasible and would not affect the other prime contractors. The general contractor used the same state-of-the-art scheduling tools that the construction manager did and had demonstrated in the past that it was capable of planning and executing recovery efforts.
Could we demonstrate to the owner that the recovery schedule is achievable?	We focused on key elements of the recovery schedule to demonstrate to the owner that the plan was achievable. For instance we outlined specific intermediate milestones such as "all BAF concrete work must be done by mid-December;" "working through the winter will be required for the finish work;" and "BAF gallery ceiling must be done with prefabricated concrete beams."
Could the recovery schedule be "proved out" through development of a detailed integrated CPM schedule?	It was easy to demonstrate on paper that the plan was achievable. This was achieved through backward scheduling from the completion date and then verifying that the resources were available to implement the plan.
Could the recovery schedule be developed that maintained an acceptable level of float time to account for future unforeseen conditions?	This was the most problematic issue. Almost any form of recovery results in the adverse effect of limiting float time. Float time is valuable because it provides schedule contingency for unforeseen conditions. It is also hard to determine in advance what an acceptable level of float might be. In retrospect, at Metro, the recovery schedule probably did not maintain an acceptable level of float time. The level of float in the recovery schedule, however, was not any less than what was contained in the original master CPM.

TABLE 5.1 (*Continued*)

Recovery Feasibility Analysis

Issue	Analysis
Could a recovery schedule be implemented that did not negatively affect the owner's current operations?	Because the recovery plan maintained the already established ACJ major milestones, this was not an issue.
Are there enough available resources to implement all aspects of the recovery schedule?	During construction of Metro there was a planned $2.2 billion development directly adjacent to the site. Had this project begun concurrently with work on Metro there would have been a potential for impact on the availability of labor, materials, and equipment. In the end, it was decided that this actually favored implementation of the recovery schedule because without it, the 13-month delay would almost certainly cause an overlap with the development activities.
Could appropriate control mechanisms be established to monitor adherence to the recovery schedule?	The construction manager and each of the prime contractors had demonstrated real-time controls for schedule monitoring. The integrated Primavera Software allowed for real-time input of construction monitoring data and an accurate current forecast of both intermediate target dates and the ACJ major milestones.

professionals, the inspection team, and contractors have added diligence in applying the established quality control and assurance procedures.

In the end, the team agreed that recovery was the best choice as it was shown to be the least expensive option, and the costs could be reimbursed by the funding agencies. The time-is-money argument prevailed, and the recovery schedule was implemented in July 2002. Although our analysis clearly showed this was the best option, it was by far the most difficult choice because it required the team to take a proactive approach to the situation. Change orders had to be negotiated and issued to the contractors to pay them for their planned recovery efforts. The team also had to feel confident that the contractors could achieve recovery and not negatively affect project quality. We went into it with our eyes wide open (see Figure 5.11).

In July 2002, prior to the start of the recovery efforts, the schedule had shown a delay in the major milestone of 13 months. The schedule projected the achievement of the ammonia removal limits in June 2005 instead of the ACJ major milestone of May 2004. Through the implementation of the recovery schedule, the actual date for achievement of the

TABLE 5.2

Cost and Benefits of Recovery

Consideration	Analysis
How much would it cost to recover?	These costs included working extended shifts (overtime, weekend, and holiday work), labor inefficiencies (overlap of trades and fatigue), additional supervision, and working during winter conditions. Costs also included several "value engineering" suggestions to save schedule time, including revising the cast-in-place concrete to precast for the BAF gallery ceiling and using PVC instead of steel conduit for the embedded electrical items.
	The cost for implementation of the recovery schedule was negotiated with the general contractor in two separate change orders. Total cost of recovery for Contract 4A was $2.98 million. When added to the recovery change order for Contract 3, total cost for all recovery efforts was $3.13 million.
	The change orders were carefully written to include stipulations that all contract provisions, including the assessment of liquidated damages for late completion, were still applicable. In essence, the owner, through execution of the change orders, "bought the completion dates" and the contractor accepted all liability in event that the recovery effort was unable to mitigate the delays. The change orders also clearly stated that they fully reimbursed the contractor for all schedule impacts that occurred prior to the recovery schedule date of July 31, 2002.
The negative impact on quality.	Sometimes quality can be negatively affected during recovery efforts, especially if work is accelerated or allowed to proceed in adverse weather conditions.
Benefits of Recovery	
Eliminating the cost of delay.	We developed a cost estimate and it showed that if the project were allowed to be delayed by 13 months the cost impact would be $8.03 million. Cost included extended overhead for each of the prime contractors, owner, construction manager, and design professional; ACJ penalties for late completion; extension of time-related unit price items such as dewatering and treatment of the dewatering effluent; and escalation.

TABLE 5.2 (*Continued*)

Cost and Benefits of Recovery

Consideration	Analysis
Good will associated with completing the project on time.	As a public project, there was a certain amount of good will in completion of the project by the dates established in the ACJ. The program was designed to improve the quality of water in Onondaga Lake. Metro was designed to contribute significantly to that by reducing the ammonia and phosphorus levels of the effluent going into the lake.
Cost for recovery would be reimbursed by the funding agencies.	The project was funded through various state and federal agencies. Cost to implement the recovery was predetermined to be fundable because it provided a reasonable and feasible approach to maintain the major milestones established in the ACJ.

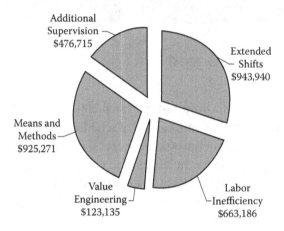

FIGURE 5.9
Cost of recovery: $3,132,247.

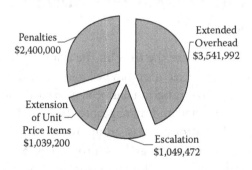

FIGURE 5.10
Cost of delay: $8,030,664.

FIGURE 5.11
Metro interior (BAF toward the HRFS).

ammonia removal limits occurred in March 2004, 15 months earlier than originally forecast and 2 months ahead of schedule. And the recovery plan cost $4.9 million less than it would have cost if the project were permitted to be delayed by 13 months. In addition to completing ahead of schedule and for less cost, the project was built to a high standard of quality. To date (early 2013), the entire complex systems are functioning successfully.

5.4 CHAPTER SUMMARY AND KEY IDEAS

5.4.1 Chapter Summary

The execution phase is where the tires hit the road. From a management perspective, construction execution can be looked at as simply the process of managing change. If the program has been governed correctly from the start, change will be limited and manageable. Tools and techniques from the monitoring and control process group are used to manage change within each project, component, and the program as a whole. Change is managed by establishing a baseline, monitoring adherence to the plan, and taking corrective action as required.

5.4.2 Key Ideas

1. All construction programs are executed in three discrete phases: design, procurement, and construction. Proper monitoring and control at each phase in the process ensures that projects remain aligned with both tactical and strategic goals and that ultimately the program produces its expected benefits and value.

2. During execution the design stage can be considered a project itself as it produces a specific outcome in a defined period of time. As such, in addition to solving the technical challenges of the projects and program, the team also monitors and controls the budget and schedule gates of conceptual design, design development, and construction documents. The techniques of value engineering and constructability analysis can be used to take corrective action if the project or program goes off target during the design phase. Reduction of scope should only be used as an option of last resort.

3. Regardless of the project delivery method, the procurement process must be open and fair. The project delivery method will determine the structure for team accountability and set the rules of engagement. When procuring the services of a design professional it is important to understand the difference between delivery risk and professional responsibility. A design professional is required by law to protect the public's health, safety, and welfare, therefore qualifications, and not price, should be the primary consideration. When procuring the services of a contractor the choice generally is made based on the bidder with the lowest price that is responsive and responsible. A responsive bidder has fulfilled the requirements of the bidding process. A responsible bidder can perform the work as specified.

4. During execution it is critical to engage the contractor as soon as possible as properly integrating its work with the rest of the team is a principle component of a successful construction project. A formal partnering process is one proven way to make that happen. Partnering is a way to encourage trust, open communication, commitment, flexible attitudes, and win–win solutions.

5. In construction, managing the master schedule is the most critical of the monitoring and controlling processes. It is critical to integrate the contractor's work plan with the master schedule to make it an accurate baseline for monitoring progress. During the integration process, the role of the program team is to work with the contractor

to make the contractor's plan work in the context of the master schedule. The attitude should be "trust with verification."

6. During planning the master budget is developed as a comprehensive register of all program budgets categorized as indirect costs, direct expenses, and contingencies. Different approaches are required to monitor and control direct versus indirect costs. For most construction programs the majority of costs in the indirect category will be associated with the support of the program management office. The largest type of controllable indirect costs is staffing expenses. Staffing expenses extend the entire duration of the program and thus the focus should be on the rate of expenditure over time. For direct costs, which are directly attributable to the design and physical work of the projects and components, the focus should be on change management.

7. In order to realize the full benefits of a program the component projects must be built with quality and also contribute to the value of the program. During the execution phase the project teams focus on construction quality and the program team focuses on value and benefits delivery. The PMI summarizes it best: "Program managers ensure and check alignment; project managers keep the detail of each project under control" [6].

8. The case study of the Metro project demonstrates the time-tested adage that time is money. Schedule impacts will almost always result in additional cost to the owner and contractor. Because of these additional costs, as managers we should look for proactive and cost-effective ways to mitigate schedule impacts. Proper execution of recovery schedules is one effective way to ensure projects and programs are completed on time and within budget. The successful implementation of a recovery schedule requires the following steps:

- The need for a recovery schedule was effectively established.
- The feasibility of the recovery schedule was verified.
- A cost–benefit analysis of implementing the recovery schedule was performed.
- The project team committed to the recovery schedule.
- The recovery schedule was effectively monitored using real-time controls to ensure conformance to the plan.

6

The Closure Process

6.1 INTRODUCTION

A successful program is closed when it has realized its full potential of benefits and value. The closure process ensures the program, and its component projects, are properly terminated through the formal acceptance of the results. Once closed, a program must transition to operational status. And once transitioned, a plan to sustain the benefits and value should be put in place. The final step is to reflect with the team on lessons learned and ways to improve future performance.

It has been my experience that the closure process is seldom fully implemented on construction programs. All too often the team is dismantled long before formal closure occurs. And, unfortunately, in the construction business being the "last one on the job" often can be an indication of poor personal performance as the "star players" have already moved on to their next assignments. Although this is short-sighted from a management perspective, it is, nonetheless, sometimes the reality, and it can have a direct impact on the effectiveness and efficiency of the closure process.

6.2 ACCEPTANCE OF THE RESULTS

The common construction term for the closure process is *closeout*. The goal of the closeout process is the formal acceptance of the results. When done properly this will include the release of the construction contractor from contractual obligations, the smooth transition of the project to operational status, and in most cases the incremental realization of program benefits. A successful closeout means different things to each member of the team. To the contractor it means resolving the punch list and collecting the final payment. To the design professional it is a project that functions as intended. To the construction professional it is a project that is delivered on time, within budget, and to a high standard of quality. And to the program team it is a project that contributes its anticipated value to the program.

During the planning process the rules of engagement for the closeout process are established in the program management plan, the general requirements, and the special provisions to the construction contracts. Comparable to the other process groups, closure is managed through monitoring and controlling the plan and taking corrective action as required.

6.2.1 Managing Contract Closeout

If not correctly planned or executed, the closeout of the construction contracts can be challenging and frustrating. This is ironic because both parties of the contract (owner and contractor) have ample incentive to close out the contracts properly and expedite final acceptance of the project. In most cases properly fulfilling the obligations of closeout is a prerequisite to the final release of the contractor's *retainage*. Retainage is a portion of a contract payment that is held until full performance of the contract terms. In construction, retainage will typically represent 5%–10% of the construction contract price, which in many cases will be more than the profit the contractor will make on the project. So from the contractor's perspective a smooth closeout process can determine whether the project will be a financial success. From the owner's perspective a well-managed closeout process will ensure that they "get what they paid for" and a smooth transition from construction to operations.

There are many standards processed, often mandated for public work, for closing out construction contracts. During the planning process the team will establish programwide closeout specifications and include them in the general requirements or special provisions of the construction contracts. Examples of effective closeout specifications are included in Chapter 4. The main focus during the execution of the closeout process should be to ensure project quality, cross-project integration, and the integrity of the program. A secondary focus should be on expediting the process as it is in everyone's interest to do so. Regardless of the project delivery method or program type, the closeout process will include the six basic steps shown in Figure 6.1.

I often tell my clients that we have no "silver bullet" to make the closeout process easy or less frustrating. Unfortunately, through experience, I have learned to set low expectations. Although the steps are well defined,

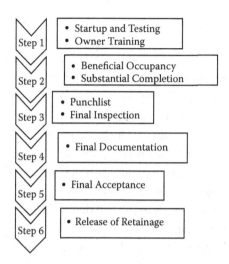

Step 1	• Startup and Testing • Owner Training
Step 2	• Beneficial Occupancy • Substantial Completion
Step 3	• Punchlist • Final Inspection
Step 4	• Final Documentation
Step 5	• Final Acceptance
Step 6	• Release of Retainage

FIGURE 6.1
Standard closeout process.

implementation of the process is seldom smooth. There are many reasons for this, but primarily it comes down to timing. Often the energy and focus for this critical task are just not there at the end of the project.

There are things that can be done to increase the chances of success. Having a clear standardized process that is well communicated and has the team's commitment is critical. Because money is a prime motivator for the contractors including specific closeout activities in the contractor's schedule of values in addition to calling attention to retainage is effective. For the same reason, allowing the contractor to close out portions of the project in advance, and releasing partial retainage, is effective. Having the right project staff, with adequate knowledge and time to manage closeout effectively, is important.

6.2.2 Transitioning to Operations

Whether it is a roadway, a wastewater treatment plant, a new high school, or the renovation of an historic building, all projects and programs cannot provide their full value without the proper transition from construction to operational status. In construction we refer to this process as *project commissioning*. Project commissioning is a quality-focused process that ensures all systems and components are designed, installed, tested, operated, and maintained according to the operational requirements of the owner. Commissioning activities are applicable to all phases of the project, from planning, design, procurement, construction, and the final handover to the owner.

During planning, *commissioning activities* focus on assuring the owner's operational requirements are properly integrated in the construction documents. This will include the details of systems tests and procedures. The technical specifications will also include performance criteria, maintenance requirements, and owner training. During construction the emphasis is on monitoring and controlling the process with the focus on coordination, witnessing, and verification. And depending on the nature of the program, commissioning activities sometimes includes an assisted operation phase as well.

Commissioning requires the active participation of the entire team. The program team will be responsible for planning, defining commissioning procedures, coordination, and quality assurance and control. The construction contractor is typically tasked with executing commissioning tests and inspections. On larger or more complex projects, it may even be appropriate to employ a commissioning authority whose sole role is to oversee the commissioning process. In fact, this is a current mandate for all LEED certified facilities.

If managed properly and done correctly, a formal commissioning process can provide many benefits. This is especially true for complex mechanical facilities and high-performance (LEED) buildings. For these types of programs the commissioning process has been proven to optimize energy use and reduce operational costs [53]. Commissioning also ensures the owner's operational and maintenance (O&M) staff is properly oriented and trained. Commissioning also improves record keeping and the documentation of installed building systems.

6.2.3 Reporting Lessons Learned

In the construction business experience is everything. There are things in construction that simply cannot be learned without living through the challenge of managing a complex project or program. I like to tell my staff and students that truly to learn from experience, "you have to have done it wrong at least once in the past and then never want to experience doing it wrong again." My guidance to my clients is that "the value of an experienced construction manager is that it is hoped they have made most of their mistakes already." Learning from experience does not, however, happen automatically. We must proactively capture the benefit of learning from both the anguishes and celebrations. I think the poet Mary Oliver puts it best in her instructions for living: "Pay attention-Be astonished-Tell about it." The process of preparing for, and then writing, a final lessons learned report is a great way to do all three.

The main focus of a lessons learned report is to reflect on what went right, what went wrong, and to explore areas for improvement. It is a collection of the team's thoughts, ideas, and notes with the sole purpose of learning from mistakes and successes. It should be a collaborative effort, but must be championed, and led, by the program manager to be successful. This will show management's commitment and support for the process.

Writing, and then sharing, a lessons learned report has many benefits for both the team and the organizations involved in the program, including:

- *Cost Reduction.* Mistakes on construction programs are often expensive and in most cases can be easily prevented. Discovering, and then sharing, the best approach to a similar problem or issue will reduce mistakes and will automatically lead to cost savings.
- *Efficiency Gains.* Formally documenting lessons learned and then centrally storing them where they are accessible organizationwide will save both time and effort during the next initiation and planning processes. This lessons learned database becomes the "organizational knowledge" that can be passed from one program to the next.
- *Continuous Improvement.* By constantly optimizing performance from one program to the next, and then distributing that knowledge, the team and organization will continuously improve.

To be most effective a lessons learned report should be comprehensive, written in a uniform format, and be concise. Most organizations will have a standard format for lessons learned reporting that will make sharing of the information more efficient and beneficial.

6.3 CHAPTER SUMMARY AND KEY IDEAS

6.3.1 Chapter Summary

The closure process ensures the program and its component projects are properly terminated through the formal acceptance of the results. Once closed, a program must transition to operational status. Once a program is transitioned a plan to sustain the benefits and value must be put in place. The final step is to reflect with the team on lessons learned and ways to improve.

6.3.2 Key Ideas

1. Closeout is the common construction term for the project closure process. There are many standardized or mandated processes for closeout. Because it is in everyone's interest to expedite the closeout process, the program team should focus on supporting the contractor in properly and expeditiously reaching final acceptance and release of retainage.

2. In order to obtain the full value of a program it must be successfully transitioned from construction to operational status. In construction this process is referred to as commissioning. In buildings, a properly executed commissioning process has been proven to optimize energy use and reduce operational costs. Commissioning will also ensure adequate training and orientation of the owner's O&M staff and improve the final documentation of results.

3. The final act of any program should be reporting on lessons learned. There are many benefits to preparing and writing a final program lessons learned report including long-term cost savings and efficiency gains through continuous process improvement. Because the process is so valuable, but seldom properly implemented, organizations should have a standard mandated process for lessons learned reporting.

Endnotes

1. Project Management Institute. 2004. *A Guide to the Project Management Body of Knowledge (PMBOK Guide)*. 3rd ed., Newtown Square, PA: Project Management Institute, Inc. p. viii, 388 pp.
2. Merriam-Webster Inc. 2003. *Merriam-Webster's Collegiate Dictionary*. 11th ed. Springfield, MA: Merriam-Webster. 1623 pp.
3. Rosenberg, J.M. 1978. *Dictionary of Business and Management*. New York: Wiley, xii, 564 pp.
4. PMI's program management credential (PgMP) originated in 2007.
5. CMAA Webiste http://cmaanet.org/about-cmaa
6. Safari Tech Books Online and Project Management Institute. 2008. *The Standard for Program Management*. Newtown Square, PA: Project Management Institute. p. xxvi, 324 pp.
7. The Construction Management Association of America. 2011. *An Owners Guide to Construction and Program Management,* 24 pp.
8. US Congress, House Committee on Government Reform. 2005. Digging up the facts: Inspecting the Big Dig and the performance of federal and state government in providing oversight of federal funds: Hearing before the Committee on Government Reform, House of Representatives, 109th Congress, first session, April 22, 2005. Washington, DC: US GPO. For sale by the Supt. of Docs., US GPO iii, 132 pp.
9. *Boston Globe*, 2008. "Big Dig's Red Ink Engulfs State." July 17.
10. Kats, G. 2006. *The Greening of America's Schools - Costs and Benefits*. The United States Green Building Council. 26 pp.
11. *Construction Data Market Intelligence*. 2011. RS Means Reed, May.
12. Gramer, J.G. 1997. The decline and fall of the SSC. *Analog Science Fiction and Fact Managazine*.
13. Bizony, P. 2009. *One Giant Leap:Apollo 11 Forty Years On*. London: Aurum Press. 158 pp.
14. Owen, H. 2004. "What Makes a Leader." *USA Today*.
15. Goleman, D. 1995. *Emotional Intelligence*. New York: Bantam. p. xiv, 352 pp.
16. Minnick, W.C. 1968. *The Art of Persuasion*. 2d ed. Boston: Houghton Mifflin. p. vi, 295 pp.
17. Buchanan, J.M., R.A. Musgrave, and ebrary Inc. 1999. *Public Finance and Public Choice Two Contrasting Visions of the State*. Cambridge, MA; London: MIT Press. p. vii, 272 pp.
18. CREATE Program Final Feasiblity Plan. 2011. Amendment 1, January.
19. Note: Data from the Association of American Railroads.
20. Note: Data from the US Department of Transporation.
21. Note: Considered by the US National Weather Service as the second worst blizzard of the 20th century.
22. Note: Extracted from an article in the *New York Times*, "Freight Train Late, Blame Chicago." May 2012.
23. Note: ABC's *Extreme Makeover Home Edition* is a reality television series in the United States that provides free home improvements for needy families.

24. Fondahl, J.W. 1962. *A Non-Computer Approach to the Critical Path Method for the Construction Industry.* 2d ed. Stanford, CA: Stanford University, Dept. of Civil Engineering. 85 pp.
25. Doyle, A.C. and Morley, C. 1930. *The Complete Sherlock Holmes.* Garden City, NY: Doubleday. 2 v., 1,122 pp.
26. Note: The hydrodemolition process uses high-pressure water to remove deteriorated concrete selectively from bridge decks.
27. Real Estate Investment Center. 2010. *Construction Survey,* Syracuse, NY.
28. Kendrick, T. 2009. *Identifying and Managing Project Risk Essential Tools for Failure-Proofing Your Project.* New York: AMACON. p. viii, 360 pp.
29. Note: SEQR (State Environmental Quality Review) is a mandated process in New York State to determine if a program will have an adverse effect on the environment.
30. Rollins, S.C. and Lanza, R.B. 2004. *Essential Project Investment Governance and Reporting: Preventing Project Fraud and Ensuring Sarbanes-Oxley Compliance.* Boca Raton, FL: J. Ross. p. xxiii, 263 pp.
31. Note: Bridging is a form of the design–build alternative project delivery method. Bridging includes the addtional services of a project criteria consultant. The project criteria consultant develops the plans and performance specifications to the conceptual level for the purpose of obtaining more competitive bids from the design–builder.
32. MacCollum, D.V. 1995. *Construction Safety Planning.* New York: Van Nostrand Reinhold. p. xii, 285 pp.
33. Note: Industrial fatalities do not include deaths from heat, pneumonia, heart trouble, etc.
34. Note: Data from the OSHA website: www.osha.gov
35. Note: Spearin V. U.S. (248 U.S. 132 (1918)), 135–136.
36. Note: Transactions of the Federal Construction Council, National Research Center (US). 1982.
37. Note: Special project provisions were developed over several years and several projects and programs, essentially taking the best parts from each experience. Many of the concepts and provisions were also extracted from my work reviewing and then revising the LIPO specifications (early 2000) to conform to the agency CM management approach.
38. Note: A material safety data sheet (MSDS) is a written document that outlines information and procedures for handling and working with chemicals.
39. Note: The American Society for Testing and Materials (ASTM) sets standards for materials, products, and services.
40. Note: American Association of State Highway and Transportation Officials (AASHTO) beams are made from precast concrete.
41. Website: http://www.state.nj.us/transportation/caphical phases
42. Note: A typical wick drain is approximately 4 inches wide, 1/8 inch thick, and 1,000 feet in length. They are used for accelerating the settlement rate for compressible soils.
43. Note: Data from the South Jersey Transportation Authority.
44. Miles, L.D. 1961. *Techniques of Value Analysis and Engineering.* New York: McGraw-Hill. 267 pp.
45. *Municipal Sewer and Water Magazine.* 2012. "A Better Way Forward." September, p. 10.
46. Cone, L. 2004. *Project Management Tools.* March.
47. Parkinson, N.C. 1955. "Parkinson's Law," *The Economist.* November 19.

48. Pacific Northwest Laboratory. 2003. "Characterizing Building Construction Decision Process." October.

49. Bresnen, M. and Marshall, N. 2001. *Partnering in Construction*, Boca Raton, FL: Taylor & Francis, October.

50. Ronco, W. and J. 1996. *Partnering Manual for Design and Construction*. New York: McGraw-Hill.

51. Bloom, M. 1997. "Partnering, a Better Way of Doing Business," Systems Engineering Process Office, MITRE.

52. Augustine, N.R. 1997. *Augustine's Laws*. 6th ed. Reston, VA: American Institute of Aeronautics and Astronautics. p. xv, 365 pp.

53. National Institute of Building Science. 2012. "Building Commissioning by the Whole Building Design Guide Project Management Committee." June 11.

Index

Page numbers followed by f indicate figure
Page numbers followed by t indicate table

O

Occupational Safety and Health Act
    (OSHA), 25, 83
Onondaga Lake, 2, 8, 14, 15, 118, 144
On-Site Safety Representative, 101
Operations, transitioning to, 163–164
Opportunities, in identifying risk, 77,
    81–82
Opportunity cost, 48
Original Baseline Schedule, 96
OSHA, 83, 84, 116
OSHA requirements for site-specific plans
    accident/incident investigation
        procedures, 86
    emergency action plan, 86–87
    record keeping, 86
    safety responsibility, 85
    safety rules, 85
    safety training, 86
    site access and control, 85
    worksite analysis, 85
Outsourcing, 75

P

Pacesetting style leadership, 43
Panama Canal, 82
Parametric cost estimate, 69–70, 70t
Partnering and project labor agreements,
    131–133, 159
Payment schedule, 108–110
PCCP, *see* Pre-stressed concrete cylinder
    pipe (PCCP)
Performance grading, of contractors,
    74–75
Performance metrics, 139
Permitting, in payment schedule, 110
Peter Kiewit Sons, Inc., 9, 49, 104, 105
PgMO, 67, 140
Planned value, *see* Budgeted Cost of Work
    Scheduled (BCWS)
Planning and design work package, in
    payment schedule, 110–111
Planning process, 23–33, 35; *see also*
        Program management plan
    case study of effective, 108–114
    introduction to, 57
Planning process group, 3, 5–6t

Planning requirements, 96–98
PLAs, *see* Project labor agreements
    (PLAs)
PMBOK guide
    basics of, 2–3
    knowledge areas, 3–4, 5–6t
    project management process, 3, 5–6t
PMI, 1, 2, 3, 134
    process-based management
        approach, 22
    program mandate, 44
    Standard for program
        management, 10
    vs. CMMA, 7–8
PMP, *see* Project Management
    Professional (PMP)
Polyvinylchloride (PVC) conduit, 32
Poor Richard's Almanac, 57
Post-bid conference, 130
Pre-bid conference, in construction
        procurement phase, 127–128
Pre-bid RFI management, in
        construction procurement
        phase, 128
Preconstruction tasks, 24
Preconstruction workshop, 131
Predecessor task, 64
Preliminary CPM Schedule, 97, 135
Preprogram preparations, 16
Prereferendum phase, 23
Pre-stressed concrete cylinder pipe
        (PCCP), 145, 146
Primal Leadership (Goleman), 42
Primavera Contract Manager, 91
Private sector, CM-at risk in, 27
Probabilistic analysis, 79
Procedure, setting, 35, 39
Process
    defined, 3
    establishing, 35, 39
Process-based management approach
    case study of, 17–18
    PMIs, 22
Procurement, 125–130
Procurement documents, development of,
    32–33
Procurement management, 4, 6t
Procurement schedule, 24
Program mandate, PMI and, 44

Printed in the United States
by Baker & Taylor Publisher Services